Lecture Notes in Mathemati

Editors:
A. Dold, Heidelberg
F. Takens, Groningen

Springer
Berlin
Heidelberg
New York
Barcelona
Budapest
Hong Kong
London
Milan
Paris
Santa Clara
Singapore
Tokyo

Hiroaki Aikawa   Matts Essén

Potential Theory – Selected Topics

Springer

Authors

Hiroaki Aikawwa
Department of Mathematics
Shimane University
Matsue 690, Japan
E-mail: haikawa@riko.shimane-u.ac.jp

Matts Essén
Department of Mathematics
Uppsala University
Box 480
75106 Uppsala, Sweden
E-mail: matts.essen@math.uu.se

Cataloging-in-Publication Data applied for

Die Deutsche Bibliothek - CIP-Einheitsaufnahme

Aikawa, Hiroaki:
Potential theory : selected topics / Hiroaki Aikawa ; Matts
Essén. – Berlin ; Heidelberg ; New York ; Barcelona ; Budapest
; Hong Kong ; London ; Milan ; Paris ; Santa Clara ; Singapore
; Tokyo : Springer, 1996
  (Lecture notes in mathematics ; 1633)
  ISBN 3-540-61583-0
NE: Essén, Matts:; GT

Mathematics Subject Classification (1991): 31B05, 31A05, 31B15, 31B25

ISSN 0075-8434
ISBN 3-540-61583-0 Springer-Verlag Berlin Heidelberg New York

Typesetting: Camera-ready TEX output by the authors
SPIN: 10479829   46/3142-543210 - Printed on acid-free paper

During the academic years 1992–1994, there was a lot of activity on potential theory at the Department of Mathematics at Uppsala University. The main series of lectures were as follows:

- A An introduction to potential theory and a survey of minimal thinness and rarefiedness. (M. Essén)
- B Potential theory. (H. Aikawa)
- C Analytic capacity. (V. Eiderman)
- D Lectures on a paper of L.-I. Hedberg [24]. (M. Essén)
- E Harmonic measures on fractals (A. Volberg)

These lecture notes contain the lecture series A,B and references for C. The E lectures will appear as department report UUDM 1994:32: Zoltan Balogh, Irina Popovici and Alexander Volberg, Conformally maximal polynomial-like dynamics and invariant harmonic measure (to appear, Ergodic Theory and Dynamical Systems).

H. Aikawa spent the Spring semester 1993 in Uppsala. V. Eiderman spent the Spring semesters 1993 and 1994 here. A. Volberg was in Uppsala during May 1994. In addition to giving excellent series of lectures, our visitors were also very active participants in the mathematical life of the department.

Uppsala September 21, 1994

Matts Essén

Contents

POTENTIAL THEORY PART I

Matts Essén

1. Preface

The first part of these notes were written to prepare the audience for lectures by H. Aikawa on recent developments in potential theory and by A. Volberg on harmonic measure. It was assumed that the participants were familiar with the theory of integration, distributions and with basic functional analysis.

Section 2 to 8 give an introduction to potential theory based on the books [7] and [29]. We begin by discussing two definitions of capacity. In the first case, the capacity of a set is the supremum of the mass which can be supported by the set if the potential is at most one on the set (cf. Section 5; also the remarks at the end of Section 8.3). This definition is taken from L. Carleson's book [7]: the book deals with general kernels, takes us quickly to interesting problems but is short on details. In the second case, the inverse of the capacity of a set is the infimum of the energy integral if the total mass is one (cf. Section 8.1). This definition is taken from the book of Landkof [29]. Landkof considers α–potentials and the corresponding α–capacity and gives many details.

When discussing α–potentials in Section 8, we can often use results from the general theory in the first six sections: it applies without change in the case $0 < \alpha \leq 2$. When $2 < \alpha < N$, where N is the dimension of the space, it is no longer possible to use the definition in Section 5, since there is no strong maximum principle in this case (cf. Theorem 4.3 and the remarks preceding Theorem 5.11). We can always use results which do not depend on the strong maximum principle: we have therefore tried to make clear what can be proved without applying this result.

In Sections 9 – 16, there is a survey of minimal thinness and rarefiedness. Minimal thinness of a set E at infinity in a half-space is defined in terms of properties of the regularized reduced function $\hat{R}_h^E$ of a minimal harmonic function with pole at infinity (cf. Section 11). A set E is defined to be rarefied at infinity in a half-space D if certain positive superharmonic functions in D dominate $|x|$ on E (cf. Definition 12.4). Characterizations of these exceptional sets in terms of conditions of Wiener type involving Greeen energy and Green mass appear here as Theorems 11.3 and 13.1.

In [2], Aikawa uses singular integral techniques to study problems in potential theory. In Section 14, we consider Green potentials and Poisson integrals in a half-space and carry through the program of Aikawa for these kinds of potentials.

In Section 15, we show that Green capacity and Green mass can in the Wiener-type conditions be replaced by ordinary capacity. This is one of the starting points of the work of Aikawa on "quasi-additivity" of capacity (cf. Section 16). There are also other interesting consequences (see the remark at the end of Section 15).

It was a pleasure to give these lectures. The participants were very active and our discussions led to many improvements in the final version of these notes. I am particularly grateful to Torbjörn Lundh for typing these notes using the $\mathcal{A}\mathcal{M}\mathcal{S}$-LATEX--system. There has been a lot of interaction between us, many preliminary versions have been circulated in the class and Torbjörn has with enthusiasm done a tremendous amount of work.

These lecture notes are dedicated to the memory of Howard Jackson of McMaster University who died in 1986 at the age of 52. Together, we tried to understand exceptional sets of the type discussed in Sections 11-15.

2. Introduction

Let us start with some "hard analysis" following Carleson's book [7]. Let our universe be $\mathbb{R}^N$, unless otherwise specified. We begin with a discussion of a family of sets that we will use.

2.1. Analytic sets. Analytic sets can be seen as a generalization of Borel sets. A situation where analytic sets arise is when you have a Borel set of a product space $X \times Y$ and project this set on X. This new subset of X is not necessarily a Borel set but it is always an analytic set.

Analytic sets have two properties in common with Borel sets: they are closed under countable intersections and countable unions, but unlike Borel sets, analytic sets are not closed under complementation.

A Borel set is always analytic, but an analytic set does not have to be a Borel set. If the complement of an analytic set is analytic then it is a Borel set.

One property of analytic sets which is of importance in connection with capacities is that for every finite Borel measure μ and every analytic set A we have that the outer measure of A is equal to the inner measure of A, i.e.

$$\mu^*(A) := \inf_{\mathcal{O} \supset A} \mu(\mathcal{O}) = \sup_{F \subset A} \mu(F) =: \mu_*(A).$$

Here the $\mathcal{O}$ and F denote open and closed sets, respectively. This convention is often used in these notes. For a thorough discussion of analytic sets we refer to [8] or [10, Appendix I].

2.2. Capacity. Let f be a non-negative set function, defined on compact sets such that $f(F_1) \leq f(F_2)$ if $F_1 \subset F_2$, where F_1, F_2 are compact.

For E bounded we define,

(1) $f(E) := \sup_{F \subset E} f(F)$, F compact. (f is an interior "measure"[1].)

It follows that

(2) $f(E_1) \leq f(E_2)$ if $E_1 \subset E_2$.

Furthermore, we assume that

(3) $f^*(E) = \lim_{n \to \infty} f^*(E_n)$, $E_n \nearrow E$, where f^* is the outer "measure" defined by $f^*(E) := \inf_{\mathcal{O} \supset E} f(\mathcal{O})$, $\mathcal{O}$ open.

The following definitions can now be made,

DEFINITION 2.1. *Capacity is a set function satisfying the above conditions 1,2 and 3.*

DEFINITION 2.2. *A set E is capacitable (with respect to f) if $f^*(E) = f(E)$.*

THEOREM 2.3 ([7] PAGE 3). *If compact sets are capacitable, then all analytic sets are capacitable.*

As an example, let us study the following set function:

$$f(E) = \begin{cases} 1 & \text{if } E^\circ \neq \emptyset, \\ 0 & \text{if } E^\circ = \emptyset. \end{cases}$$

We now choose $E = \{0\}$, and will have $f^*(E) = 1$ but $f(E) = 0$; telling us that E is not capacitable with respect to f.

---

[1] f need not be additive.

2.3. Hausdorff measures. Let h be an increasing, continuous function from $[0, \infty)$ to $[0, \infty)$ and assume further that $h(0) = 0$. The classical choice is $h(r) = r^s, s > 0$. If E is a bounded set in $\mathbb{R}^N$ we can cover it by a sequence of balls, $\{B_\nu\}$, where $B_\nu = B(x_\nu, r_\nu)$, i.e. a ball in $\mathbb{R}^N$ centered at x_ν with radius r_ν. Having $E \subset \bigcup B_\nu$ we define,

DEFINITION 2.4 ($M_h(E)$). $M_h(E) = \inf \sum h(r_\nu)$, taken over all such coverings of E.

With an extra condition on the size of the r_ν's we also state,

DEFINITION 2.5 ($\Lambda^{(\rho)}(E)$). $\Lambda^{(\rho)}(E) = \inf \sum h(r_\nu)$, taken over all coverings of E such that, $E \subset \bigcup B_\nu$ and $\sup r_\nu \leq \rho$.

We see that $\Lambda^{(\rho)}(E)$ increases as ρ decreases to 0. This will finally lead to

DEFINITION 2.6 (THE CLASSICAL HAUSDORFF MEASURE).
$$\Lambda_h(E) = \lim_{\rho \to 0} \Lambda^{(\rho)}(E).$$

While studying Hausdorff measures it is perhaps worthwhile mentioning Hausdorff dimension, which is a very important concept in the theory of fractals. If we let $h(r) = r^s$ then $\Lambda_s(E) := \Lambda_h(E)$ is the outer s-dimensional Hausdorff measure (restricted to Borel sets E). We note that $\Lambda_s(E)$ decreases as s increases. There exists a s_0 such that,
$$\Lambda_s(E) = \begin{cases} \infty & \text{if } s \in (0, s_0), \\ 0 & \text{if } s > s_0. \end{cases}$$
s_0 is called the Hausdorff dimension of E, i.e. $s_0 = \dim(E)$. The Hausdorff dimension coincides with the Euclidean dimension for the cases when $s_0 \in \mathbb{N}$; e.g. $\dim(\text{line}) = 1$, $\dim(\text{plane}) = 2$ and so on.

A non trivial example is the $\frac{1}{3}$–Cantor set, which has Hausdorff dimension $\frac{\log(2)}{\log(3)} \approx 0.6309$. See [18] for further details on this subject. It is not always suitable to cover our set E by balls; sometimes a net will do the work better , or at least differently. To give a good definition of a net we first need to define the notion of a cube.

DEFINITION 2.7 (CUBE). A set of the form $\{x \in \mathbb{R}^N : a_i \leq x_i \leq b_i\}$ where $b_i - a_i$ is constant over the indices i.

DEFINITION 2.8 (NET). A net is a division of $\mathbb{R}^N$ into cubes $\{Q\}$ all of the same side-length L with sides parallel to the coordinate axis, such that the cube $\{x \in \mathbb{R}^N : 0 \leq x_i \leq L\}$ is in the net. Moreover we have $\bigcup \bar{Q} = \mathbb{R}^N$ and $Q_i^0 \cap Q_j^0 = \emptyset$ if $i \neq j$.

The dyadic refinement is a useful way to construct nets. Let G_p be a net with $L = 2^{-p}$. The next generation, G_{p+1} is obtained from G_p by dividing every cube in G_p into 2^N subcubes each of side-length 2^{-p-1}. We can now form the family of all dyadic cubes, $G = \{G_p\}_{p=-\infty}^\infty$.

Consider now a set E covered by $\bigcup Q_\nu, Q_\nu \in G$. The side-length of Q_ν is δ_ν and we define:

DEFINITION 2.9. $m_h(E) = \inf \sum_\nu h(\delta_\nu)$, for all such coverings $\bigcup Q_\nu$ of E.

Remark
- It does not matter if the balls $\{B_\nu\}$, which were used in the definition of M_h, are closed or open.
- It is a convention to let the cubes $\{Q_\nu\}$ be closed.

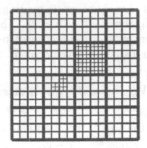

FIGURE 2.1. A dyadic net under construction.

LEMMA 2.10. *There are constants, C_1 and C_2 dependent only on the dimension, N, such that*

$$C_1 M_h(E) \le m_h(E) \le C_2 M_h(E).$$

PROOF. (of the last inequality.) Let us first consider the planar case $(N = 2)$. To cover a ball of radius r by cubes, or in our case squares, we need at most 5×5 squares of side-length 2^{-p} when $2^{-p} \le r \le 2^{-p+1}$. See figure 2.2 for the geometrical argument. If we now pick a sufficiently good ball covering of E, i.e. $\sum h(r_\nu) \le M_h(E) + \epsilon$, then we have

$$m_h(E) \le 25 \sum h(r_\nu) \le 25 M_h(E) + 25\epsilon.$$

So we conclude that C_2 can be chosen as 25 in the two–dimensional case. For $N > 2$ we just exchange 25 for 5^N. The other inequality is treated in an analogous way. $\square$

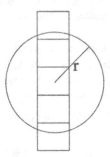

FIGURE 2.2. To cover a ball by squares.

THEOREM 2.11 (FROSTMAN 1935). *Let μ be a non–negative and sub–additive set function such that*

(1) $\mu(B) \le h(r)$ *for every ball of radius r,*

then

(2) $\mu(E) \le M_h(E).$

Conversely, there exists a constant a, depending only on the dimension, such that for every compact set F, there exists a measure μ such that

$$\mu(F) \ge a M_h(F)$$

and μ satisfies equation (1).

PROOF. First, we show the easy part, $(1) \Rightarrow (2)$. If $E \subset \bigcup B_\nu$ then by covering by balls we have $\mu(E) \leq \sum \mu(B_\nu) \leq \sum h(r_\nu)$, the last inequality is just condition (1). Now we can take infimum over such coverings and get $\mu(E) \leq \inf \sum h(r_\nu)$, which is (2).

The second part of the theorem will be proved for the planar case ($N = 2$). Fix a large integer n and consider a net G_n, see figure (2.3). Define a measure, μ_n to have

FIGURE 2.3. F is captured in a net, G_n.

constant density on each $Q \in G_n$ and the value:

$$\mu_n(Q) = \begin{cases} h(2^{-n}) & \text{if } Q \cap F \neq \emptyset, \\ 0 & \text{if } Q \cap F = \emptyset. \end{cases}$$

Let us look at a lower level in the net, a larger square Q in G_{n-1}. We will then have two possibilities: either $\mu_n(Q) \leq h(2^{-n+1})$ or $\mu_n(Q) > h(2^{-n+1})$. In the latter case we change the measure by multiplying by a constant times the characteristic function of Q, i.e. $c\chi_Q$, where c is defined by $c\mu_n(Q) = h(2^{-n+1})$. This is done for every Q in G_{n-1} and the resulting measure is called μ_{n-1}.

Continuing this way "down" in the net, we arrive after n steps at the measure μ_0. We do know that the relation $\mu_n(Q) \leq h(2^{-\nu})$ holds for every Q in G_ν for every ν in $[0, n]$. Since our resulting μ_0 depends on the starting level n, we indicate this by writing $\mu_0^{(n)}$ instead of μ_0. It is then possible to find a weakly converging subsequence of the sequence $(\mu_0^{(n)})_{n=1}^\infty$, i.e.

$$\mu_0^{(n_k)} \rightharpoonup \mu.$$

It only remains to check that the resulting measure μ has the right properties. It is fairly easy to see that $supp\ \mu \subset F$ and $\mu(Q) \leq h(2^{-\nu}), \forall Q \in G_\nu$, telling us that condition (1) is valid. Furthermore, let $\epsilon > 0$ be given, we can then choose a covering $\{Q_j\}, Q_j \in G$ so that $F \subset \bigcup Q_j$ and $m_h(F) + \epsilon \geq \sum h(\delta_j)$. Now, let n be large. For each $Q \in \{Q_j\}$ with side-length $= \delta$, we have two possibilities,

$$\text{either } \mu_0^{(n)}(Q) = h(\delta) \text{ or } \overset{*}{\sum} \mu_0^{(n)}(Q_k) = h(\delta).$$

(The $*$ indicates summation over all $Q_k \subset Q$.) The total mass of $\mu_0^{(n)}$ is then bounded below:

$$\mu_0^{(n)}(\mathbb{R}^N) \geq \inf \overset{*}{\sum} h(\delta_j).$$

The same inequality holds for the limit measure μ. That is,

$$\mu(\mathbb{R}^N) \geq \inf \overset{*}{\sum} h(\delta_j),$$

and it follows that

$$\mu(\mathbb{R}^N) \geq m_h(F).$$

By the fact that $supp\ \mu \subset F$ one concludes that $\mu(F) \geq m_h(F)$. Using lemma 2.10 we finally observe

$$\mu(F) \geq m_h(F) \geq a M_h(E),$$

ending the proof. $\square$

Claim. $M_h(E)$ is an outer measure. That is, $M_h(E) = \inf_{\mathcal{O} \supset E} M_h(\mathcal{O})$, $\mathcal{O}$ is open.

This is easy to see. Given an $\epsilon > 0$ there is a collection of (open) balls, $\{B_\nu\}$ such that $\mathcal{O} = \bigcup B_\nu$, $\mathcal{O} \supset E$ and $\sum h(r_\nu) \leq M_h(E) + \epsilon$ giving,

$$M_h(\mathcal{O}) + \epsilon \geq M_h(E) + \epsilon \geq \sum h(r_\nu) \geq M_h(\mathcal{O}).$$

Thus, $M_h(E)$ is an outer measure. What about $m_h(E)$? It turns out that $m_h(E) = \inf_{\mathcal{O} \supset E} m_h(\mathcal{O})$ is true when $N = 1$, but not necessarily for $N \geq 2$. The problem in higher dimensions occurs due to the fact that it is necessary to add many more cubes than we can control by an estimate of the contributed terms in the defining sum of $m_h(E)$.

The following lemma is closely related to the above questions.

LEMMA 2.12 ($N = 1$). $\lim_{n \to \infty} m_h(E_n) = m_h(E)$, $E_n \nearrow E$.

PROOF. Take a sequence $\{\epsilon_n\}$ to be defined later and let $\{\omega_{\nu n}\}$ be closed dyadic intervals that cover E_n such that $\sum_\nu h(\delta_{\nu n}) \leq m_h(E_n) + \epsilon_n$. Then we pick, for every $x \in E$, the largest interval in the sequence, $\{\omega_{\nu n}\}$, containing x and call it ω. We may also assume that $\frac{k}{2^m}$ is not in E for any $m, k \in \mathbb{Z}$ (just remove a countable set). The different "non–intersecting" intervals ω —countably many— that we obtain this way, are denoted $\{\omega_\mu\}$ and have lengths $\{\delta_\mu\}$.

An alternative way of constructing the sequence $\{\omega_\mu\}$ is the following; if $\omega', \omega'' \in \{\omega_{\nu n}\}$ and $\omega' \subset \omega''$, then throw away ω'. Keep on until we have denumerably many intervals ω_μ such that,

$$\bigcup_{n=1}^{\infty} \bigcup_{\nu=1}^{\infty} \omega_{\nu n} = \bigcup \omega_\mu.$$

One also notes that ω_μ^o and $\widetilde{\omega_\mu}^o$ are disjoint if $\omega_\mu \neq \widetilde{\omega_\mu}$. We have obtained a new sequence by throwing away the smaller, already covered, sets. Obviously $E_n \subset \bigcup \omega_\mu$.

Choose an integer n and consider those ω_μ's that are taken from $\{\omega_{\nu 1}\}$. They cover a certain subset Q_1 of E_n. The same subset is covered by a certain subsequence of $\{\omega_{\nu n}\}$ denoted $\{\omega_{\nu n}\}^{(1)}$ which is a sequence of subintervals of the chosen ω_ν's. We now claim that

$$\sum^{(1)} h(\delta_\mu) \leq \sum^{(1)} h(\delta_{\nu n}) + \epsilon_1.$$

To show this we assume the contrary and observe

$$\sum h(\delta_{\nu 1}) = \sum^{(1)} h(\delta_\mu) + (\text{"the rest"}),$$

where "the rest" could be estimated from below by $m_h(E_1 \setminus Q_1)$. By the assumption, we know that

$$\sum^{(1)} h(\delta_\mu) > \sum^{(1)} h(\delta_{\nu n}) + \epsilon_1.$$

Since $\{\omega_{\nu n}\}^{(1)}$ covers Q_1 and therefore the smaller set $Q_1 \bigcap E_1$ we have the estimate

$$\sum^{(1)} h(\delta_{\nu n}) \geq m_h(Q_1 \bigcap E_1).$$

We obtain, after putting it all together,

$$\sum h(\delta_{\nu 1}) > m_h(Q_1 \bigcap E_1) + \epsilon_1 + m_h(E_1 \setminus Q_1) = \epsilon_1 + m_h(E_1),$$

which is the desired contradiction.

The next step is to consider the ω_ν's taken from $\{\omega_{\nu 2}\}$ but not from $\{\omega_{\nu 1}\}$. Again, we find $\sum^{(2)} h(\delta_\mu) \leq \sum^{(2)} h(\delta_{\nu n}) + \epsilon_2$. Repeat this argument for all coverings $\{\omega_{\nu k}\}_{k=1}^n$! We obtain n inequalities that can be added, giving

$$\sum_{k=1}^n \sum^{(k)} h(\delta_\mu) \leq \sum h(\delta_{\nu n}) + \sum_{\nu=1}^n \epsilon_\nu \leq m_h(E_n) + \sum_{\nu=1}^n \epsilon_\nu + \epsilon_n.$$

If we now let n tend to infinity we will have

$$m_h(E) \leq \sum_1^\infty h(\delta_\mu) \leq \lim_{n \to \infty} m_h(E_n) + \sum_{\nu=1}^\infty \epsilon_\nu.$$

Since $\sum_{\nu=1}^\infty \epsilon_\nu$ can be chosen arbitrarily small we find

$$m_h(E) \leq \lim_{n \to \infty} m_h(E_n)$$

Trivially, we also have $m_h(E_n) \leq m_h(E)$ concluding the proof. $\square$

Remark The same argument holds for $N > 1$.

2.4. Is m_h a capacity? Let us now specialize to the case when E_n is compact and E is open and bounded. Lemma 2.12 gives us then

(3) $$m_h(\mathcal{O}) = \sup_{F \subset \mathcal{O}} m_h(F)$$

where F is compact and $\mathcal{O}$ is open and bounded.

What about the other relation, the outer relation?

(4) $$m_h(E) = \inf_{\mathcal{O} \supset E} m_h(\mathcal{O}), \quad \mathcal{O} \text{ open},$$

where E is arbitrary. We know it is true for $N = 1$; but what happens otherwise?

EXERCISE 2.1. *Find sufficient conditions on h so that equation (4) holds for each set E in $\mathbb{R}^N$. What can we say when $h(r) = r^{N-\alpha}, N \geq 2$, or when $h(r) = (\log^+ \frac{1}{r})^{-1}, N = 2$?*

Let us repeat the conditions for the set function f to be a capacity.

(1) $f(E) := \sup f(F)$ for all compact F that are subsets of E.
(2) $f(E_1) \leq f(E_2)$ if $E_1 \subset E_2$
(3) $f^*(E) = \lim_{n \to \infty} f^*(E_n)$, where $E_n \nearrow E$.

So, is $f_0 := m_h$ a capacity? We have only to check the last condition, the first is clear due to lemma 2.12 and equation (3) and the second condition follows immediately from the definition of m_h.

Given $E_n \nearrow E$ find nested open sets $\mathcal{O}_n$ such that $\mathcal{O}_n \supset E_n$ and $f_0^*(\mathcal{O}_n) \leq f_0^*(E_n) + \epsilon$. Thus $E = \bigcup E_n \subset \bigcup \mathcal{O}_n$. We then have $f_0^*(E) \leq f_0^*(\bigcup \mathcal{O}_n)$ but since $\bigcup \mathcal{O}_n$ is open $f_0^*(\bigcup \mathcal{O}_n) = f_0(\bigcup \mathcal{O}_n)$. Using lemma 2.12 gives us $f_0(\bigcup \mathcal{O}_n) = \lim_{n \to \infty} f_0(\mathcal{O}_n)$ but this is bounded above by the assumptions on $\mathcal{O}_n$ by $\lim_{n \to \infty} f_0^*(E_n) + \epsilon$, giving us

$f_0^*(E) \leq \lim_{n \to \infty} f_0^*(E_n)$ The opposite inequality is trivial. Hence the answer is yes, $f_0 := m_h$ is a capacity.

We turn now, impatiently, to the next question: "Are compact sets capacitable?" Or in other words: "Is $f^*(F) = f(F)$?" or, equivalently :"Does (4) hold?" The equivalence follows immediately from the definitions of capacity $f^*(F) = \inf_{\mathcal{O} \supset F} m_h(\mathcal{O})$ and $f(F) = m_h(F)$.

Remark: We can modify m_h by slightly changing the meaning of "cover". Let $\{Q_\nu\}$ "cover" a set E if every $x \in E$ is an interior point of $\bigcup Q_\nu$. Then we define,

$$m_h'(E) := \inf \sum h(\delta_\nu),$$

where infimum is taken over all such coverings.

EXERCISE 2.2. *Is equation (4) true with m_h replaced by m_h'?*
Is $\lim m_h'(E_n) = m_h'(E)$ *as* $E_n \nearrow E$?

3. The Physical background of Potential theory

Behind the theory there are old and interesting questions about the physical reality around us. The classical examples are the theories of gravity and electrostatics. More information on this can be found in [35].

3.1. Electrostatics in space. Consider an electrically charged body, say negatively charged. If we now look at a test particle nearby the body, we will find that there is a force acting on the particle due to the presence of the charged body, see picture 3.1. The force, $\vec{F}$, on the test-charge is $\vec{F} = e\vec{\mathcal{E}}(x, y, z)$, where $\vec{\mathcal{E}}(x, y, z)$ is

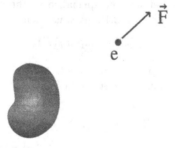

FIGURE 3.1. A test particle, e, in an electric field

a vector field generated by the charged body. If we consider the special case where the body is just a point–charge, with charge $e > 0$, at the location $\bar{x} \in \mathbb{R}^3$, then the electric field is given by the law of Coulomb's.

$$\vec{\mathcal{E}}(x) = \frac{e}{r^2} \frac{x - \bar{x}}{r} = \frac{e}{r^3}(x - \bar{x}).$$

At the end of the 18th century it was observed by Lagrange that there exists a scalar function, φ, with

$$\vec{\mathcal{E}} = -\nabla\varphi,$$

where $\varphi(x) = \frac{e}{|x - \bar{x}|}$.

If we have several charges, with charge e_i at x_i the scalar function becomes $\varphi(x) = \sum \frac{e_i}{|x - x_i|}$. The electrical field will still be $\vec{\mathcal{E}}(x) = -\nabla\varphi(x)$. Suppose now we have charges continuously distributed over a body, rather than a finite set of point charges. That is, suppose there is a continuous function ρ defined on the body Ω such that for every portion B of Ω, the charge in B is given by $\int_B \rho\, dV$. Then, $\varphi = \int_\Omega \frac{\rho(\xi)\, dV}{|x-\xi|}$ for all $x \in \mathbb{R}^3 \setminus \Omega$ and as before, $\vec{\mathcal{E}} = -\nabla\varphi$. The Laplacian can, in distributional sense, be computed

$$\Delta\varphi = -4\pi\rho.$$

The scalar function φ is the potential function of $\vec{\mathcal{E}}$ and these functions will be studied in the next section.

4. Potential theory

We need the fundamental solution of Laplace's equation

$$\Phi(x) = \phi(r) = \begin{cases} \log\frac{1}{r} & \text{if } N = 2 \\ r^{2-N} & \text{if } N \geq 3. \end{cases}$$

Here $r = |x|$ and $\Delta\Phi = -c_N\delta$ with the constant $c_N > 0$. Let now $H : \mathbb{R} \to [0, \infty)$ be a continuous, increasing and convex function. (The convexity is not always essential and this condition will later be removed in some cases.) We shall study kernels of the form

$$K(r) = H(\phi(r))$$

and we will also assume integrability

$$\int_0 K(r) r^{N-1}\, dr < \infty.$$

We allow ourselves to write $K(x) = K(|x|)$ letting us write the above condition

$$\int_{|x|<a} K(x)\, dx < \infty.$$

With respect to $K(r)$, we form the *potential* of the set function σ

$$u_\sigma(x) := \int K(|x - y|)\, d\sigma(y),$$

and the *energy integral*

$$I(\sigma) := \iint K(|x - y|)\, d\sigma(y)\, d\sigma(x)$$

or, using the newly defined potential,

$$I(\sigma) = \int u_\sigma(x)\, d\sigma(x).$$

The potential and the energy of course have physical origins; see [35].

Remarks:

- There is a problem when $N = 2$ because $\log\frac{1}{|z|}$ changes sign , where $z \in \mathbb{C}$. This will be studied later.
- If we allow $K(r) = r^{\alpha-N}$ we will get something called an α–potential.

Let us look at an example of "strange" behavior of a potential in $\mathbb{R}^N$. Let $u(x) = \sum \frac{a_i}{|x-y_i|}$ where $\{y_i\}$ is a dense set in $\mathbb{R}^N, a_i > 0, \sum a_i < \infty$. We have then $u \in L^1_{loc}$ but $u(y_i) = \infty$! What are the sets $\{x : u(x) > A\}$?

From now on we will restrict ourselves to non–negative measures, μ, with bounded support. We have then the following lemma on the property of semi–continuity. First, the definition,

DEFINITION 4.1 (SEMI–CONTINUITY FROM BELOW). *A function, f, is semi–continuous from below[2] if there exists an increasing sequence of continuous functions such that $\lim_{n\to\infty} f_n(x) = f(x)$ everywhere.*

EXERCISE 4.1. *Prove that if f is semi–continuous from below then*

  (i) $\underline{\lim}_{x\to x_0} f(x) \geq f(x_0)$,
  (ii) *the sets $\{x : f(x) > A\}$ are open for every A.*

LEMMA 4.2. *Let μ be a measure with compact support. Then*

$$(a) \qquad \liminf_{x\to x_0} u_\mu(x) \geq u_\mu(x_0).$$

If $\mu_n \rightharpoonup \mu$ weakly then

$$(b) \qquad \liminf_{n\to\infty} u_{\mu_n}(x) \geq u_\mu(x),$$

$$(c) \qquad \liminf_{n\to\infty} I(\mu_n) \geq I(\mu).$$

PROOF. Let us introduce the cut–off function, $K_n = min(K, n)$, which is a continuous and bounded function.

(a)   Define

$$u_n(x) := \int K_n(|x - y|)\, d\mu(y).$$

Clearly, $u_n(x) \nearrow u_\mu(x)$ as $n \to \infty$. Let us assume that $u_\mu(x_0)$ is finite. For ϵ given and n large enough, we have $u_n(x_0) > u_\mu(x_0) - \epsilon$. Thus, in a neighbourhood $\mathcal{O}_n$ of x_0, we have

$$u_\mu(x) \geq u_n(x) \geq u_n(x_0) - \epsilon \geq u_\mu(x_0) - 2\epsilon.$$

We conclude that

$$\liminf_{x\to x_0} u_\mu(x) \geq u_\mu(x_0) - 2\epsilon$$

and (a) follows. If $u_\mu(x_0) = \infty$, a slightly different argument works.

(b)   By the weak convergence

$$\int K_p(|x - y|)\, d\mu_n(y) \to \int K_p(|x - y|)\, d\mu(y) \text{ as } n \to \infty.$$

But, since

$$\int K_p(|x - y|)\, d\mu_n(y) \leq u_{\mu_n}(x)$$

and

$$\int K_p(|x - y|)\, d\mu(y) \to u_\mu(x)$$

as p tends to infinity, we have

$$\underline{\lim}_{n\to\infty} u_{\mu_n}(x) \geq u_\mu(x).$$

(c)   The weak convergence is still valid in $\mathbb{R}^N \times \mathbb{R}^N$, i. e.

$$d\mu_n(x)\, d\mu_n(y) \rightharpoonup d\mu(x)\, d\mu(y).$$

Thus we have

$$\iint K_p(|x - y|)\, d\mu_n(y)\, d\mu_n(y) \to \iint K_p(|x - y|)\, d\mu(y)\, d\mu(y) \text{ as } n \to \infty$$

and we can proceed as in the proof of (b). $\square$

---

[2]or, lower semi-continuous.

4.1. The maximum principle.

THEOREM 4.3 (THE (STRONG) MAXIMUM PRINCIPLE). *If H is as in the introduction and if $u_\mu(x) \leq 1$ on the support S_μ of μ, then $u_\mu(x) \leq 1$ everywhere.*

Remark. In this proof, we must assume that H is convex.

PROOF. Egorov's theorem is applicable (details are given at the end of this section). Given an $\epsilon > 0$ there exists a closed set F such that F is a subset of S_μ,

$$\mu(F) \geq \mu(S_\mu) - \epsilon$$

and $u_\mu(x)$ converges uniformly on F. Let us define a new measure, μ_1 as the restriction of μ to F, i.e.

$$\mu_1(e) := \mu(e \cap F).$$

This measure will also generate a potential that is uniformly convergent on F, i.e. for any $x_0 \in F$,

$$(5) \qquad \int_{|y-x_0|<4\eta} K(|y-x_0|)\,d\mu_1(y) < \delta,$$

where η only depends on δ.

Let $\{x_n\}$ be a sequence tending to $x_0 \in F$. Let $\eta > 0$ be given. Then

$$\limsup_{n\to\infty} u_{\mu_1}(x_n) \leq \limsup_{n\to\infty} \int_{|y-x_n|\geq\eta} K(|y-x_n|)\,d\mu_1(y) +$$

$$+ \limsup_{n\to\infty} \int_{|y-x_n|<\eta} K(|y-x_n|)\,d\mu_1(y).$$

The first term is, due to continuity and boundedness, simply

$$\int_{|y-x_0|\geq\eta} K(|y-x_0|)\,d\mu_1(y).$$

The second term has to be treated more gently. From now on, we consider only values of n which are such that $|x_n - x_0| < \eta$. We divide the analysis in two parts:

(1) If $x_n \in F$, then we use equation (5) obtaining,

$$\int_{|y-x_n|<\eta} K(|y-x_n|)\,d\mu_1(y) \leq \int_{|y-x_n|<4\eta} K(|y-x_n|)\,d\mu_1(y) < \delta.$$

(2) If $x_n \notin F$, then things are more intricate. Let $\Gamma_{\nu n}$ be a cone with x_n at its vertex and let $\xi_{\nu n} \in F \cap \Gamma_{\nu n}$ be the point that is closest to x_n, or one of them if there is not a unique one, see figure 4.1. Assume that the diameter of the

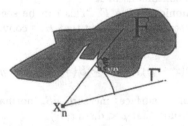

FIGURE 4.1. The cone $\Gamma_{\nu n}$.

spherical cap that $\xi_{\nu n}$ lies in is less than $\frac{1}{2}|x_n - \xi_{\nu n}|$. Then we have for any $y \in F \cap \Gamma_{\nu n}$ that $|y - \xi_{\nu n}| \leq |y - x_n|$ Now, take a number of such cones, how many depends on the dimension N, that cover $\mathbb{R}^N$. Assume we need M such cones. We have then for the sequence $\{\xi_{\nu n}\}_{\nu=1}^M$ the following estimate for all $y \in F$,

$$\int_{B(x_n, \eta)} K(|y - x_n|)\, d\mu_1(y) \le \sum_1^M \int_{B(x_n, \eta) \cap \Gamma_{\nu n}} K(|y - \xi_{\nu n}|)\, d\mu_1(y),$$

where $B(x_n, \eta) = \{y : |y - x_n| < \eta\}$.

These integrals have also to be treated for two cases.

(a) If $|x_0 - \xi_{\nu n}| < 2\eta$, we have, $K(|y - x_n|) \le K(|y - \xi_{\nu n}|)$, if $y \in \Gamma_\nu \cap F$. We can then estimate,

$$\int_{B(x_n, \eta) \cap \Gamma_{\nu n}} K(|y - x_n|)\, d\mu_1(y) \le$$

$$\le \int_{B(x_n, \eta)} K(|y - \xi_{\nu n}|)\, d\mu_1(y) \le$$

$$\le \int_{B(\xi_{\nu n}, 4\eta)} K(|y - \xi_{\nu n}|)\, d\mu_1(y).$$

Hence, using equation (5),

$$\int_{|y - x_n| < \eta} K(|y - x_n|)\, d\mu_1(y) \le$$

$$\le \sum_1^M \int_{|y - \xi_{\nu n}| < 4\eta} K(|y - \xi_{\nu n}|)\, d\mu_1(y) < M\delta.$$

(b) If $|x_0 - \xi_{\nu n}| > 2\eta$, then the set, $F \cap B(x_0, \eta) \cap \Gamma_{\nu n}$ is empty, giving,

$$\int_{|y - x_n| < \eta} K(|y - x_n|)\, d\mu_1(y) \le$$

$$\le \sum_1^M \int_{B(x_n, \eta) \cap \Gamma_{\nu n}} K(|y - \xi_{\nu n}|)\, d\mu_1(y) = 0.$$

Hence,

$$\limsup_{n \to \infty} u_{\mu_1}(x_n) \le u_{\mu_1}(x_0).$$

Due to the lower semi–continuity we conclude,

$$\lim_{x \to x_0} u_{\mu_1}(x) = u_{\mu_1}(x_0),$$

for all $x_0 \in \mathbb{R}^N$. i.e. u_{μ_1} is continuous in $\mathbb{R}^N$.

Claim. u_{μ_1} is subharmonic outside F. This can be seen by the use of Jensen's inequality, which can be used due to the fact that H is convex, where H is defined by $H(\Phi(x)) = K(x)$.

$$\int_{|y| = \rho} H(\Phi(x + y))\, dS_\rho(y) \ge H\left(\int_{|y| = \rho} \Phi(x + y)\, dS_\rho(y)\right),$$

with the $(N-1)$–dimensional "surface" measure, dS_ρ, normalized on the set $\{y : |y| = \rho\}$. We have chosen the ρ such that $\rho < \text{dist.}(x, F)$.

Since Φ is harmonic outside F we have the meanvalue property, giving,

$$H\left(\int_{|y| = \rho} \Phi(x + y)\, dS_\rho(y)\right) = H(\Phi(x)).$$

Thus, u_{μ_1} fulfills the demands on a subharmonic function outside F. Using this fact and the maximum principle for subharmonic functions we have,

$$u_{\mu_1}(x) \leq 1, \forall x \in \mathbb{R}^N,$$

since,

$$u_{\mu_1}(x) \begin{cases} \leq 1 & \text{if } x \in S_\mu \supset F. \\ \to 0 & \text{as } |x| \text{ tends to } \infty. \\ \leq 1 & \text{on } \partial F. \end{cases}$$

At last, we go back to the original measure. Let us pick a $x \notin S_\mu$ and let ρ be the distance between x and S_μ. We can then estimate,

$$u_\mu(x) \leq u_{\mu_1}(x) + K(\rho)\epsilon \leq 1 + \epsilon K(\rho).$$

Since ϵ can be chosen arbitrarily small, we are finished. $\square$

THEOREM 4.4 (THE CONTINUITY PRINCIPLE). *If $u_\mu(x)$ is continuous on S_μ, then it is continuous everywhere.*

PROOF. Let as usual $u_n(x) := \int K_n(x-y)\,d\mu(y)$. Then the continuous $u_n(x)$ tends to the continuous $u_\mu(x)$ as n tends to ∞. Dini's theorem gives us that $u_n(x)$ converges uniformly on the support S_μ, i.e. for every $\epsilon > 0$ and every $x \in S_\mu$ there exist N_0 only depending on ϵ such that, for all $\eta > 0$,

$$\int_{B(x,\eta)} K(x-y)\,d\mu(y) < \epsilon + \int_{B(x,\eta)} K_n(x-y)\,d\mu(y),$$

for all $n \geq N_0$. Furthermore,

$$\int_{B(x,\eta)} K_n(x-y)\,d\mu(y) \leq n \int_{B(x,\eta)} d\mu(y),$$

which can be estimated,

$$n \int_{B(x,\eta)} d\mu(y) \leq \frac{n}{K(\eta)} \int_{B(x,\eta)} K(x-y)\,d\mu(y) \leq \frac{n}{K(\eta)} \max_{x \in S_\mu} u_\mu(x) \leq n\frac{\epsilon}{n},$$

if we choose η small enough.

Combining the terms, we obtain,

$$\int_{B(x,\eta)} K(x-y)\,d\mu(y) < 2\epsilon.$$

Hence we can take $\mu_1 = \mu$ in the previous proof and thus u_{μ_1} is continuous everywhere. $\square$

In the proof of Theorem 4.3, we used Egorov's theorem. To explain this more in detail, we take the cut–off function K_n in use again and consider

$$u_n(x) = \int K_n(x-y)\,d\mu(y).$$

The sequence $\{u_n\}$ converges everywhere to $u_\mu(x)$ on S_μ. By Egorov's theorem, there exists for each $\delta > 0$ a set E_δ with $\mu(E_\delta) < \delta$ such that $\{u_n\}$ converges uniformly to u_μ on the set $S_\mu \setminus E_\delta$. We claim that u_n converges uniformly to u_μ on the closure F of $S_\mu \setminus E_\delta$. In fact, if $\{x_j\}$ is a sequence in $S_\mu \setminus E_\delta$ converging to $x_0 \in F$, then

$$\int \Big(K(x_0-y) - K_n(x_0-y) \Big)\,d\mu(y) \leq$$

$$\leq \liminf_{j \to \infty} \int \Big(K(x_j-y) - K_n(x_j-y) \Big)\,d\mu(y) < \epsilon, \text{ if } n > N(\epsilon).$$

For each n, we choose η_n so that $K(\eta_n) = 2n$. If $x_0 \in F$, then

$$n\mu(|y - x_0| < \eta_n) \leq \int_{S_\mu} \left(K(x_0 - y) - K_n(x_0 - y) \right) d\mu(y) < \epsilon,$$

if $n > N(\epsilon)$.

We conclude that uniformly on F, we have

$$(6) \qquad \int_{|y-x_0|<\eta} K(x - y) \, d\mu(y) \leq \epsilon + \int_{|y-x_0|<\eta} K_n(x - y) \, d\mu(y) \leq$$
$$\leq \epsilon + n\mu(|y - x_0| < \eta) < 2\epsilon,$$

if $\eta < \eta_n$ and $n > N(\epsilon)$. The choice of η depends only on ϵ.

4.2. α–potentials. As an example of a family of potentials we have the α–potentials defined as,

$$u_\alpha^\sigma(x) := \int \frac{d\sigma(y)}{|x - y|^{N-\alpha}}.$$

A natural question to ask is whether the maximum principle is applicable or not? To be able to use the previous theorem we have to check when the function H defined by $r^{\alpha-N} = H(r^{2-N})$ is convex. We limit the discussion to the case $N > 2$. By changing the variable we obtain the following condition to get a convex H: is $H(t) = t^{\frac{N-\alpha}{N-2}}$ convex? This is equivalent to the condition $\frac{N-\alpha}{N-2} \geq 1$ or $\alpha \leq 2$.

What about the case $\alpha > 2$? There is then a weak form of the maximum principle shown in [34, pages 37 – 43].

THEOREM 4.5 (UGAHERI). *Let $K = H \circ \Phi$ be as above except that we do not assume H to be convex. Then there is a constant, C_0 such that if $u_\mu(x) \leq 1$ on the support S_μ then $u_\mu(x) \leq C_0$ everywhere: C_0 depends only on the dimension N and the function H.*

5. Capacity

In this section we assume that all sets are bounded. For simplicity, let us say that they are subsets of $X = \{x : |x| < M\}$. Let us define,

$$\Gamma_E := \{\mu, \ \mu \text{ pos. measure with support } S_\mu \subset E \text{ s.t. } u_\mu \leq 1, \forall x \in E\}.$$

Remark An analytic set of positive Hausdorff measure for some measure function h contains a closed subset of positive h–measure, see Chapter two in [7].

With the help of the set Γ_E we define,

DEFINITION 5.1 (CAPACITY). $C_K(E) := \sup_{\mu \in \Gamma_E} \mu(E)$,

and the analogue to the "almost everywhere" notion,

DEFINITION 5.2 (P.P.). *A property which holds except on a set of capacity zero is said to hold p.p. That is a.e. in French, "presque partout".*[3]

We first state two results on capacity.

(1) If u_ν is a bounded potential and E is an analytic set with $C_K(E) = 0$, then $\nu(E) = 0$. This will be proved later.
(2) If E is given and there is a measure μ with the three properties $\mu(X) > 0$, the support $S_\mu \subset E$ and $u_\mu(x)$ is bounded on E, then $C_K(E) > 0$.

For later use, let us present two lemmas.

---

[3]Alternatively, one can say *quasi everywhere*.

LEMMA 5.3. *If $\{F_n\}$ is a sequence of closed sets, then*

$$C_K(\bigcup_1^\infty F_n) \le \sum_1^\infty C_K(F_n).$$

PROOF. If $\epsilon > 0$ is given, we find μ with

(1) compact support S_μ contained in $\bigcup_1^\infty F_n$.
(2) $u_\mu(x) \le 1$ on $\bigcup_1^\infty F_n$.
(3) $\mu(\bigcup_1^\infty F_n) \ge C_K(\bigcup_1^\infty F_n) - \epsilon$
    If $\mu_n = \mu|_{F_n}$, we have $\mu_n \in \Gamma_{F_n}$ and
(4) $\mu(\bigcup_1^\infty F_n) \le \sum_1^\infty \mu(F_n) \le \sum_1^\infty C_K(F_n).$

(1) and (4) give the lemma. $\square$

LEMMA 5.4. *If $\{\mathcal{O}_n\}$ is a sequence of open sets, then*

$$C_K(\bigcup_1^\infty \mathcal{O}_n) \le \sum_1^\infty C_K(\mathcal{O}_n).$$

PROOF. We start as in the previous proof but replace $\bigcup_1^\infty F_n$ by $\bigcup_1^\infty \mathcal{O}_n$. To each $\mathcal{O}_n$, we find $F_n \subset \mathcal{O}_n$ such that

$$\mu(\mathcal{O}_n) \le \mu(F_n) + \epsilon/2^n, \ n = 1, 2, \ldots$$

Since $\mu_n = \mu|_{F_n} \in \Gamma_{F_n}$ and each $x \in \operatorname{supp}\mu$ is an interior point of some $\mathcal{O}_n$, we have

$$C_K(\bigcup_1^\infty \mathcal{O}_n) - \epsilon \le \mu(\mathbb{R}^N) \le \sum_1^\infty \mu(\mathcal{O}_n) \le$$

$$\le \sum_1^\infty \mu(F_n) + \epsilon \le \sum_1^\infty C_K(F_n) + \epsilon \le \sum_1^\infty C_K(\mathcal{O}_n) + \epsilon,$$

which proves the lemma. $\square$

5.1. Equilibrium distributions.

THEOREM 5.5. *Let F be a compact set and let $K(|x|)$ be any kernel. Then there exists a measure μ with $S_\mu \subset F$ such that $u_\mu(x) = 1$ p.p. on S_μ and $\mu(F) \ge C_K(F)$. If we have a strong maximum principle, then $u_\mu(x) = 1$ p.p. on F, ,$\mu \in \Gamma_F$ and $\mu(F) = C_K(F)$.*

PROOF. Let us consider the variational problem of minimizing the energy $I(\mu)$, which is a very natural problem in nature. Define $\gamma = \inf I(\mu)$ over those measures having the properties $S_\mu \subset F$ and $\mu(F) = 1$. Now, let $\{\mu_n\}$ be a sequence of "allowed" measures such that $I(\mu_n)$ tends to γ as n tends to infinity. Then we can pick a subsequence, for simplicity also called $\{\mu_n\}$, such that μ_n weakly tends to μ. Then

$$\mu(F) = \lim_{n\to\infty} \int 1 \, d\mu_n(x) = 1;$$

due to the lower semi continuity, we have

$$\gamma \le I(\mu) \le \liminf_{n\to\infty} I(\mu_n) = \gamma.$$

Thus we know that this measure μ minimizes the energy, i.e. $I(\mu) = \gamma$.
    Claim. $u_\mu(x) = \gamma$ p.p. on S_μ.

PROOF OF THE CLAIM.. We divide the proof in two parts,

(1) $u_\mu \geq \gamma$ on F except on a set of capacity zero.

Let $\epsilon > 0$ be given and assume that the compact set $T = \{x \in F : u_\mu(x) \leq \gamma - \epsilon\}$ has positive capacity. Then there exists a mass distribution τ on T with $\tau(T) = 1$ such that $u_\tau(x) \leq c_0$ for all x .

By a convex combination of the measures,

$$u_\delta := (1 - \delta)\mu + \delta\tau, \quad \text{with } \delta \in (0, 1)$$

we obtain,

$$\mu_\delta(F) = 1.$$

The energy of this combined measure is,

$$I(\mu_\delta) = \int\int K(x - y)\, d\mu_\delta(x)\, d\mu_\delta(y) =$$

$$= (1 - \delta)^2 I(\mu) + 2\delta(1 - \delta) \int u_\mu\, d\tau + \mathcal{O}(\delta^2).$$

Here we used Fubini's theorem, $\int u_\mu\, d\tau = \int u_\tau\, d\mu$.

$$I(\mu_\delta) \leq (1 - 2\delta)I(\mu) + 2\delta(\gamma - \epsilon) + \mathcal{O}(\delta^2) =$$

$$= \gamma - 2\delta\epsilon + \mathcal{O}(\delta^2) < \gamma,$$

which is the desired contradiction. Since $\{x \in F : u_\mu(x) < \gamma\} = \bigcup_1^\infty \{x \in F : u_\mu(x) < \gamma - 2^{-n}\}$, it follows from Lemma 5.3 that $u_\mu \geq \gamma$ on F except for a set of capacity zero.

(2) $u_\mu(x) \leq \gamma$ on S_μ.

The previous part gave,

$$u_\mu(x) \geq \gamma \ \text{p.p.} \ F.$$

If $u_\mu(x_0) > \gamma$ for an $x_0 \in S_\mu$, then $u_\mu(x) > \gamma$ on a neighbourhood $\mathcal{O}$ of x_0. The minimal energy,

$$\gamma = I(\mu) = \int u_\mu(x)\, d\mu(x) =$$

$$= \int_{\mathcal{O}\bigcap F} u_\mu(x)\, d\mu(x) + \int_{F\backslash\mathcal{O}} u_\mu(x)\, d\mu(x) >$$

$$> \gamma\mu(\mathcal{O}\bigcap F) + \gamma\mu(F \backslash \mathcal{O}) = \gamma.$$

This is only possible if $\mu(\mathcal{O}\bigcap F) = 0$ telling us $x_0 \notin S_\mu$, giving us the contradiction. We thus have

$$u_\mu(x) \leq \gamma \ \text{everywhere on} \ S_\mu.$$

Putting the two parts together proves the claim.

Let us now choose $\mu_0 = \mu/\gamma$. This will give us $u_{\mu_0} = 1$ p.p. on S_μ. We claim that if $\nu \in \Gamma_F$, then

$$\int u_{\mu_0}(x)\, d\nu(x) \geq \nu(F).$$

To prove this, let $\epsilon > 0$ be given. We know that the set $T = \{x \in F : u_{\mu_0}(x) \leq 1 - \epsilon\}$ is compact with $C_K(T) = 0$. It follows that $\nu(T) = 0$ since $\nu(T) = \nu|_T(T)$ where $\nu|_T \in \Gamma_T$ and $0 \leq \nu|_T(T) \leq \sup_{\mu \in \Gamma_T} \mu(T) = C_K(T) = 0$. Thus

$$\int u_{\mu_0}(x)\, d\nu(x) \geq (1 - \epsilon)\nu(F \backslash T) = (1 - \epsilon)\nu(F).$$

Since ϵ is arbitrary, the claim is proved. $\square$

Hence,

$$\mu_0(F) \geq \int u_\nu(x)\, d\mu_0(x) = \int u_{\mu_0}(x)\, d\nu(x) \geq \nu(F),$$

and,

$$C_K(F) \leq \mu_0(F),$$

which proves the first part of the theorem.

If we have a strong maximum principle, then it follows from the fact that $u_\mu(x) \leq \gamma$ on S_μ that the inequality holds everywhere. Thus $\mu_0 \in \Gamma_F$ and $u_{\mu_0} = 1$ p.p. on F. $\quad\square$

With an additional condition, we can sharpen the result above by removing the *p.p.*

THEOREM 5.6. *Assume that the strong maximum principle holds. Let F be a compact set and assume that for every $x \in F$ there exists a bounded half-cone, V_x, with vertex at x, such that $V_x \subset F$. If furthermore the "doubling condition" is valid, i.e. $\exists C$ such that $K(r) \leq CK(2r)$ when $r < \frac{1}{2}$, then there exists a measure μ in Γ_F with $u_\mu(x) \equiv 1$ on F and $\mu(F) = C_K(F)$.*

PROOF. Let us assume, without loss of generality, that $0 \in F$. Our goal is to show,

$$(7) \qquad\qquad u_\mu(0) = \gamma$$

and then use the last part of the above proof to normalize our measure.

Let us start by defining a function,

$$q(x) = \begin{cases} K(|x|)/\int_{|y|<|x|} K(|y|)\, dy & \text{if } x \in Q \text{ and } a < |x| < b \\ 0 & \text{otherwise,} \end{cases}$$

where Q is the cone, whose existence is assumed in the theorem, with vertex $0 \in F$, see figure 5.1. We see that there exists an a arbitrarily small such that $\int_{\mathbb{R}^N} q(x)\, dx = 1$

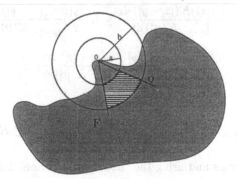

FIGURE 5.1. $q(x)$ is non-vanishing on the shaded part of the cone.

because,

$$\int_{|x|<b} \frac{K(|x|)\, dx}{\int_{|y|<|x|} K(|y|)\, dy} = \int_0^b \frac{K(r) r^{N-1}\, dr}{\int_0^r K(s) s^{N-1}\, ds} =$$

$$= \log\left[\int_0^r K(s) s^{N-1}\, ds\right]_0^b = \infty.$$

If we restrict the integral only to accept x in our cone Q, we will will still have,

$$\int_{|x|<b, x\in Q} \frac{K(|x|)\, dx}{\int_{|y|<|x|} K(|y|)\, dy} = \infty.$$

Now, since we know from the previous theorem that $u_\mu(x) = \gamma$ p.p. in F we have,

$$\gamma = \gamma \cdot 1 = \gamma \int q(x)\,dx = \int u_\mu(x)q(x)\,dx.$$

Here, we used the fact that the support of q is in F. Thus we have,

$$\gamma = \int_F d\mu(y) \int K(|x-y|)q(x)\,dx.$$

Let us now fix, $\rho > 0$, and consider two cases,

(1) $|y| > \rho$.
  Then we have,

$$\int K(|x-y|)q(x)\,dx \text{ tends to } K(|y|) \text{ uniformly as } a, b \text{ tend to } 0.$$

(2) $|y| \le \rho$.
  We have to use the "doubling condition".

$$\tilde{I} = \int K(|x-y|)q(x)\,dx = \int_{a<|x|<b, x\in Q} \frac{K(|x|)K(|x-y|)\,dx}{\int_{|t|<|x|} K(|t|)\,dt} =$$

$$= \int_{|x-y|>\frac{|y|}{2}} + \int_{|x-y|<\frac{|y|}{2}} = I_1 + I_2.$$

(a) I_1
  $|x-y| > \frac{|y|}{2}$ implies $K(|x-y|) \le K(\frac{|y|}{2})$ giving, $I_1 \le K(\frac{|y|}{2}) \cdot 1$.
(b) I_2
  $|x-y| < \frac{|y|}{2}$ implies $|x-y| < \frac{|y|}{2} \le |x|$. Hence,

$$\frac{K(|x|)K(|x-y|)\,dx}{\int_{|t|<|x|} K(|t|)\,dt} \le K(\frac{|y|}{2})\frac{K(|x-y|)\,dx}{\int_{|t|<|x|} K(|t|)\,dt} \le$$

$$K(\frac{|y|}{2})\frac{K(|x-y|)\,dx}{\int_{|t|<\frac{|y|}{2}} K(|t|)\,dt},$$

therefore,

$$I_2 \le K(\frac{|y|}{2}) \int_{x\in Q,|x-y|<\frac{|y|}{2}} \frac{K(|x-y|)\,dx}{\int_{|t|<\frac{|y|}{2}} K(|t|)\,dt} \le K(\frac{|y|}{2}) \cdot \frac{1}{1}.$$

Collecting the terms and using the "doubling condition" leads to,

$$\tilde{I} = I_1 + I_2 \le 2K(\frac{|y|}{2}) \le \text{Const.}K(|y|).$$

The contribution to the integral is,

$$\int_{|y|<\rho} d\mu(y)\tilde{I} = \int_{|y|<\rho} d\mu(y) \int K(|x-y|)q(x)\,dx \le$$

$$\le \text{Const.} \int_{|y|<\rho} K(|y|)\,d\mu(y) < \delta(\rho),$$

but $\delta(\rho)$ tends to 0 as ρ tends to 0, since we know from the proof of the previous theorem that $u_\mu(0) = \int K(|y|)\,d\mu(y) < \infty$.

We know now that the essential contribution to $\int K(|x - y|)q(x)\,dx$ comes from the case $|y| > \rho$. If we now first let $a, b \to 0$ and then $\rho \to 0$, we have,

$$\int K(|x - y|)q(x)\,dx \to K(|y|), \quad \text{uniformly,}$$

concluding,

$$\gamma = \int_F K(|y|)\,d\mu(y) = u_\mu(0),$$

and the theorem is proved. □

5.2. Three extremal problems.

THEOREM 5.7. *The extremal problems*

(A) $A^{-1} = \inf_\nu I(\nu),\ S_\nu \subset F,\ \nu(F) = 1.$
(B) $B = \inf_\nu\{\ \text{total mass of } \nu\},\ u_\nu \geq 1,\ p.p.\ \text{on } F.$
(C) $C = \sup \nu(F),\ u_\nu \leq 1\ p.p.\ \text{on } F.$

are equivalent in the sense that

$$A = B = C = C_K(F).$$

In (B) and (C), we have to assume that we have a strong maximum principle.

PROOF. Let μ be the equilibrium distribution given from Theorem 5.5. i.e.

$$\begin{cases} u_\mu(x) \geq 1 \ \text{p.p. on } F \\ \mu(F) = C_K(F) \\ \text{supp } \mu \subset F. \end{cases}$$

(A) It follows form the proof of Theorem 5.5 that any measure ν solving the (A) problem is such that $I(\nu) = 1/C_K(F)$.
(B) Assume that ν is in the B–class implying that $u_\nu \geq 1$ p.p. on F. If $\epsilon > 0$ and if $T = \{x \in F : u_\nu(x) \leq 1 - \epsilon\}$, we know that $C_K(T) = 0$ and it follows that $\mu(T) = 0$ and that

$$C_K(F) \leq \int u_\nu\,d\mu = \int u_\mu\,d\nu \leq \int d\nu$$

giving us $C_K(F) \leq B$. The opposite inequality is easy because μ "competes" in the B–class.
(C) Let ν be in the C–class, then we have $u_\nu \leq 1$ p.p. on F. If $\epsilon > 0$ and if $T = \{x \in F : u_\nu(x) > 1 + \epsilon\}$, it follows from our assumption $C_K(T) = 0$ that $\mu(T) = 0$ and that

$$C_K(F) \geq \int u_\nu\,d\mu = \int u_\mu\,d\nu \geq \nu(F),$$

giving us $C_K(F) \geq C$. Again, the opposite inequality is easy because μ competes in the C–class.

□

Remark. Let us assume that our "capacity" $C_K(\cdot)$ is such that

$$E_1 \subset E_2 \Rightarrow C_K(E_1) \leq C_K(E_2)$$

(cf. the remark before Theorem 5.11). Then, if ν is a measure and $u_\nu(x) \leq 1$ p.p. on $\text{supp}\,\nu$, then $u_\nu(x) \leq 1$ everywhere on $\text{supp}\,\nu$. To see this, we consider the set $E = \text{supp}\,\nu \cap \{x : u_\nu(x) > 1\}$ and assume that there exists $x_0 \in E$. By lower semicontinuity, there is a neighbourhood $\mathcal{O}$ of x_0 where $u_\nu(x) > 1$. Since $C_K(E) = 0$, we have $C_K(E \cap \mathcal{O}) = 0$ and we must have $\nu(E \cap \mathcal{O}) = \nu(\mathcal{O}) = 0$. It follows that $x_0 \notin$ supp ν: the contradiction shows that there is no such x_0.

THEOREM 5.8. *The solutions μ_A and μ_C of the extremal problems (A) and (C) are unique and we have $\mu_C = C_K(F)\mu_A$.*

Remark (B) need not have a unique solution. (See Exercise 5.8.)

PROOF OF THEOREM 5.8.. WLOG[4] let $K(r) \equiv 0$ for $r > r_0$, because $x, y \in F$ which is compact. The new set $F_1 := \{x - y : x, y \in F\}$ is a new compact set. Therefore we can change the definition of $K(r)$ for large values of r without changing the values of $u_\mu(x) = \int K(|x - y|)\, d\mu(y)$ for $x \in F$.

LEMMA 5.9. *For $N \geq 2$, we have*

$$F(\xi) := \int_{\mathbb{R}^N} K(x)e^{-i\langle \xi, x \rangle}\, dx > 0,$$

for all $\xi \in \mathbb{R}^N$, where $F = \hat{K}$ is the Fourier transform of K.

PROOF OF LEMMA 5.9.. Let Σ be the unit sphere, $\{x : |x| = 1\}$ and $d\sigma$ the normalized area element such that $\sigma(\Sigma) = 1$. We turn to polar coordinates and obtain,

$$F(\xi) = \int_0^\infty K(r)r^{N-1} \int_\Sigma e^{-i\langle \xi, x \rangle}\, d\sigma\, dr.$$

Furthermore, let us assume that $|\xi| = 1$. If that is not the case, just study $F(\rho\xi)$ instead, where $\rho = |\xi| \neq 1$. This change will not cause any serious trouble.

Let $x_1 = r\cos(\phi)$ then we define,

$$J_0(r) := \int_\Sigma e^{-ix_1}\, d\sigma$$

and

$$J(r) := \Re\{\int_0^{\pi/2} e^{ir\cos(\phi)}(\sin(\phi))^{N-2}\, d\phi\}.$$

We have the following relation between these two functions:

$$cJ(r) = J_0(r), \quad c > 0.$$

Furthermore, we have the following three facts,

(1) $J(0) > 0$
(2) $J'(0) = 0$
(3) $t^2 J''(t) + t(N-1)J'(t) + t^2 J(t) = 0$.

EXERCISE 5.1. *Prove the three facts above!*

We can therefore express $J(t)$ as,

$$J(t) = -J''(t) - \frac{(N-1)}{t} J'(t)$$

giving us

$$F(\xi) = c \int_0^\infty K(r)r^{N-1} J(r)\, dr =$$

$$= (-c) \int_0^\infty K(r)r^{N-1}(J''(r) + \frac{N-1}{r} J'(r))\, dr =$$

$$= c \int_0^\infty (J(0) - J(r))\, d(K'(r)r^{N-1}),$$

---

[4]Without Loss Of Generality

where we integrated by parts. We recall that $K(r) = H(\Phi(r))$. Assuming that H is convex we see that $d(K'(r)r^{N-1}) > 0$. The other factor $(J(0) - J(r))$ is also positive because,

$$(J(0) - J(r)) = \int_0^{\pi/2} (1 - \cos(\phi r))(\sin(\phi))^{N-2} d\phi > 0,$$

giving us the lemma. □

LEMMA 5.10. *For all kernels K and all signed measures $\sigma \neq 0$ with compact support such that $I(|\sigma|) < \infty$, we have $I(\sigma) > 0$.*

PROOF OF THE LEMMA.. Let $u(x) = \int K(|x - y|) d\sigma(y)$ and the energy $I(\sigma) = \int u(x) d\sigma(x)$. The Fouriertransform of σ is then $\hat{\sigma}(\xi) = \int e^{i\xi x} d\sigma(x)$. We hope to prove that this claim is true:

(8) $$I(\sigma) = \int \hat{K}(\xi)|\hat{\sigma}(\xi)|^2 d\xi$$

Proof of formula (8). Let $\Phi_n(x)$ be $\gamma_n \exp(-n|x|^2)$ normalized so that $\int \Phi_n(x) dx = 1$. Then we have $\hat{\Phi}_n > 0$ and $\hat{\Phi}_n \in L^1$. We will now use this auxiliary function to make our integrals absolutely convergent so that we can use Fubini's theorem:

$$\int \hat{\Phi}_n(\xi)\hat{K}(\xi)|\hat{\sigma}(\xi)|^2 d\xi = \int \hat{\Phi}_n(\xi)\hat{K}(\xi)\hat{\sigma}(\xi) \int e^{i\langle\xi,y\rangle} d\sigma(y) d\xi =$$

$$= \int d\sigma(y) \int \hat{\Phi}_n(\xi)\hat{K}(\xi)\hat{\sigma}(\xi)e^{i\langle\xi,y\rangle} d\xi.$$

We know that $\hat{\Phi}_n\hat{K}\hat{\sigma} \in L^1$ and

$$\Phi_n * u_\sigma(y) = \int \hat{\Phi}_n(\xi)\hat{K}(\xi)\hat{\sigma}(\xi)e^{i\langle\xi,y\rangle} d\xi, \quad \text{a.e.}$$

The a.e. can be changed to everywhere since the left-hand-side is continuous. Hence,

$$\int \hat{\Phi}_n(\xi)\hat{K}(\xi)|\hat{\sigma}(\xi)|^2 d\xi = \int \Phi_n * u_\sigma(y) d\sigma(y).$$

If we now let $n \to \infty$, then $\Phi_n \to \delta$ giving us, if u_σ is continuous then $\Phi_n * u_\sigma(y) \to u_\sigma$ pointwise. We will also have, $\hat{\Phi}_n\hat{K}|\hat{\sigma}|^2 \to \hat{K}|\hat{\sigma}|^2$ and we have proved the formula (8).

If u_σ is not continuous we have to get a good continuous approximation u_{σ_1} of u_σ. Let us divide σ into two parts, $\sigma = \sigma^+ - \sigma^-$. As in the proof of the maximum principle Theorem 4.3, we find $F \subset S_\sigma$ such that

$$|\sigma|(F) \geq |\sigma|(S_\sigma) - \epsilon,$$

where $u_\sigma(x)$ converges uniformly on F. If

$$\sigma_1(E) := \sigma(E \cap F),$$

then u_{σ_1} is continuous in $\mathbb{R}^N$ and we have,

$$(\sigma - \sigma_1)(E) = \sigma^+(E \setminus F) - \sigma^-(E \setminus F).$$

The total variation is

$$\|\sigma - \sigma_1\| = \sigma^+(\mathbb{R}^N \setminus F) + \sigma^-(\mathbb{R}^N \setminus F) = |\sigma|(S_\sigma \setminus F) < \epsilon.$$

It remains to estimate the energy difference $|I(\sigma) - I(\sigma_1)|$. Let us define a new measure $\tau := \sigma - \sigma_1$. Then we have,

$$|I(\sigma) - I(\sigma_1)| = |2\int u_\sigma d\tau - I(\tau)| \leq 2\int u_{|\sigma|} d|\tau| + \int u_{|\tau|} d|\tau|.$$

Since $|\tau| \leq |\sigma|$, we deduce that

$$2 \int u_{|\sigma|} \, d|\tau| + \int u_{|\tau|} \, d|\tau| \leq 3 \int u_{|\sigma|} \, d|\tau|.$$

We have assumed that the "total energy" $I(|\sigma|)$ is finite. It follows that

$$\int u_{|\sigma|} \, d|\tau| \to 0 \text{ as } ||\tau|| \to 0.$$

In other words,

$$0 \leq I(\sigma_1) \to I(\sigma), \text{ as } \epsilon \to 0$$

implying $I(\sigma) \geq 0$.

It remains to prove that $I(\sigma) = 0$ if and only if $\sigma = 0$. When $||\sigma_1 - \sigma|| \to 0$, we have $\hat{\sigma}_1(\xi) \to \hat{\sigma}(\xi)$ pointwise and it follows from Fatou's lemma that

$$\int_{-\infty}^{\infty} \hat{K}(\xi) |\hat{\sigma}(\xi)|^2 \, d\xi \leq I(\sigma).$$

Hence,

$$I(\sigma) = 0 \Rightarrow \int_{-\infty}^{\infty} \hat{K}(\xi) |\hat{\sigma}(\xi)|^2 \, d\xi = 0.$$

Since we know from Lemma 5.9 that $\hat{K}(\xi) > 0$, we see that $\hat{\sigma} \equiv 0$ which implies $\sigma \equiv 0$ and we have proved the lemma. $\square$

Now, we turn to the proof of the theorem. Let us first show the uniqueness of the A–problem. Let μ_1 and μ_2 be two "solutions" to the A–problem. The energy of their difference is,

$$I(\mu_1 - \mu_2) = I(\mu_1) - 2 \int u_{\mu_1} \, d\mu_2 + I(\mu_2) \leq$$

$$\leq \frac{1}{C_K(F)} - \frac{2}{C_K(F)} + \frac{1}{C_K(F)} = 0.$$

By the use of Lemma 5.10 we have $\mu_1 = \mu_2$ and we have uniqueness for the A–problem.

Considering the C–problem, let ν be a solution, i.e. $\nu(F) = C_K(F)$, let μ be a solution of the A–problem and let $\mu_1 = C_K(F)\mu$. Again, let us examine the energy of the difference,

$$I(\mu_1 - \nu) = \int u_{\nu} \, d\nu - 2 \int u_{\mu_1} \, d\nu + I(\mu_1) \leq$$

$$\leq C_K(F) - 2C_K(F) + C_K(F) = 0,$$

telling us $\mu_1 = \nu$. We have finished the proof of Theorem 5.8. $\square$

5.3. Every analytic set is capacitable. Is $C_K(\cdot)$ a capacity in the sense of Section 2.2? Let us first prove that it satisfies condition 1 there, i.e.,

$$C_K(E) = \sup_{F \subset E} C_K(F), \quad F \text{ compact.}$$

PROOF. Take an $\epsilon > 0$ and find $\mu_0 \in \Gamma_E$ such that $C_K(E) \leq \mu_0(E) + \epsilon$. Let F be the support of μ_0. Find a measure $\mu \in \Gamma_F$ with the properties $u_\mu(x) = 1$ p.p. on F and $u_\mu(x) \leq 1$ everywhere. Then we have

$$C_K(E) \geq \mu(F) \geq \mu_0(F) \geq C_K(E) - \epsilon,$$

and the proof is complete. $\square$

If we have a strong maximum principle, it is easy to see that our "capacity" $C_K(F)$ fulfills the first two conditions in Definition 2.1, and we have

$$E_1 \subset E_2 \Rightarrow \Gamma_{E_1} \subset \Gamma_{E_2}$$

which implies that $C_K(E_1) \le C_K(E_2)$.

In the case where there is only a weak maximum principle, this argument breaks down. Later, we shall introduce a different definition of capacity using the energy integral: then this difficulty will not appear for kernels of the form $K(|x|) = |x|^{\alpha-N}$, $x \in \mathbb{R}^N$, $2 < \alpha < N$. (We recall that if $0 < \alpha \le 2$, we have a strong maximum principle for such kernels).

We shall now prove that if our kernel is such that we have a strong maximum principle, our "capacity" $C_K(F)$ is in fact a capacity according to Definition 2.1. The first two conditions in the definition hold and it remains to prove

(9) $$C_K^*(E) = \lim C_K^*(E_n) \text{ as } E_n \nearrow E.$$

THEOREM 5.11. *Every analytic set is capacitable for C_K, i.e. we have $C_K(E) = C_K^*(E)$.*

We divide the proof into two parts. First, we show that all compact sets are capacitable. Secondly, we prove that (9) holds. We can the apply Theorem 2.3 which will finish the proof of Theorem 5.11.

Let F be compact and let $\mathcal{O}_n = \{x \in \mathbb{R}_N : d(x, F) < \frac{1}{n}\}$. Then we find $\mu_n \in \Gamma_{\mathcal{O}_n}$ such that $\mu_n(\mathcal{O}_n) > C_K(\mathcal{O}_n) - \epsilon_n$, $\epsilon_n \to 0$. We will in this way get a sequence of measures $\{\mu_n\}$. Let us take a weakly convergent subsequence of this sequence, which, for simplicity, will also be called $\{\mu_n\}$. We have then $\mu_n \to \mu$ and

$$u_\mu(x) \le \liminf_{n\to\infty} u_{\mu_n}(x) \le 1 \text{ on } F.$$

It follows that (cf. Exercise 5.2)

$$\mu(F) = \lim_{n\to\infty} \mu_n(\mathcal{O}_n) \ge \lim_{n\to\infty}(C_K(\mathcal{O}_n) - \epsilon_n) = \lim_{n\to\infty} C_K(\mathcal{O}_n).$$

Since $\mu \in \Gamma_F$ we have also

$$C_K(F) \ge \mu(F) \ge \lim_{n\to\infty} C_K(\mathcal{O}_n) \ge C_K^*(F) \ge C_K(F).$$

The first part is proved.

EXERCISE 5.2. *Show that*

$$\lim_{n\to\infty} \mu_n(\mathcal{O}_n) = \mu(F).$$

Let us go on with the second part of the proof.

LEMMA 5.12. *For any sets $\{E_n\}$,*

$$C_K^*(\bigcup_1^\infty E_n) \le \sum_1^\infty C_K^*(E_n).$$

PROOF. Given $\epsilon > 0$ choose open sets $\mathcal{O}_n \supset E_n$ such that $C_K(\mathcal{O}_n) \le C_K^*(E_n) + \frac{\epsilon}{2^n}$. Hence,

$$C_K^*(\bigcup_1^\infty E_n) \le C_K(\bigcup_1^\infty \mathcal{O}_n) \le$$

$$\le \sum_1^\infty C_K(\mathcal{O}_n) \le \sum_1^\infty C_K^*(E_n) + \epsilon.$$

Since ϵ is arbitrary we are finished. $\square$

LEMMA 5.13. *If $u_\mu(x) \leq 1$, then there exists, for any $\epsilon > 0$, an open set $\mathcal{O}$ such that*

(1) $C_K(\mathcal{O}) < \epsilon$.
(2) $u_\mu(x)$ *is continuous in the complement of $\mathcal{O}$* .

PROOF. Let us use the proof of Theorem 4.3 to see that for a given $\delta > 0$ there exists a restricted measure μ_1 of μ such that,

(1) $u_{\mu_1}(x)$ is continuous
    and if $u_\mu(x) = u_{\mu_1}(x) + u_{\nu_1}(x)$ then
(2) $\nu_1(X) < \delta_1$

Furthermore, $S_n := \{x : u_{\nu_1}(x) > \frac{1}{n}\}$ is an open set. Choose $F \subset S_n$ and μ_0 as an equilibrium distribution on F, then $u_{\mu_0}(x) \leq 1$ everywhere and $\mu_0(F) = C_K(F)$. Thus we have,

$$\frac{1}{n}\mu_0(F) \leq \int u_{\nu_1}\, d\mu_0 = \int u_{\mu_0}\, d\nu_1 \leq \delta_1.$$

Hence,

$$C_K(S_n) = \sup_{F \subset S_n} \mu_0(F) \leq n\delta_1.$$

Next, choose n_i and δ_i such that $n_i \to \infty$, $\delta_i \to 0$ and $\sum n_i \delta_i < \epsilon$. For each index i we get a corresponding measure ν_i as above. Let $\mathcal{O} = \bigcup S_{n_i}$. It follows from lemma 5.4 that

$$C_K(\mathcal{O}) \leq \sum C_K(S_{n_i}) < \epsilon.$$

For each n we have, $u_\mu(x) = u_{\mu_n}(x) + u_{\nu_n}(x)$ where the first term is continuous. Hence, if $x \notin \mathcal{O}$ then $x \notin S_{n_i}$ for all S_{n_i} in the union. This implies $u_{\nu_{n_i}} \leq 1/n_i$ and we have, for $x_0 \notin \mathcal{O}$,

$$\limsup_{x \to x_0, x \notin \mathcal{O}} |u_\mu(x) - u_\mu(x_0)| \leq \limsup |u_{\mu_{n_i}}(x) - u_{\mu_{n_i}}(x_0)| +$$

$$+ \limsup |u_{\nu_{n_i}}(x) - u_{\nu_{n_i}}(x_0)| \leq 0 + \limsup_{x \notin S_{n_i}} |u_{\nu_{n_i}}(x) - u_{\nu_{n_i}}(x_0)| \leq \frac{1}{n_i}.$$

That is,

$$\limsup_{x \to x_0, x \notin \mathcal{O}} |u_\mu(x) - u_\mu(x_0)| = 0,$$

since n_i can be taken as large as we like. The proof of lemma 5.13 is finished. $\square$

LEMMA 5.14. *If $\mu_n \to \mu$, then*

$$\liminf_{n \to \infty} u_{\mu_n}(x) = u_\mu(x)$$

except in a set of outer capacity 0.

PROOF. Due to lemma 5.13 there exists an open set $\mathcal{O}$ such that u_{μ_n} and u_μ are continuous outside $\mathcal{O}$ and $C_K(\mathcal{O}) < \epsilon$. Let

$$F_n := \{x : u_\mu(x) \leq r, u_{\mu_n}(x) \geq \rho, x \notin \mathcal{O}\},$$

where r and ρ are rational numbers and $r < \rho$. F_n is a closed set. We define $G_n = \bigcap_n^\infty F_k$ which is also a compact set.
    Claim. $C_K(G_n) = 0$.

If not, there exists a measure $\nu \in \Gamma_{G_n}, \nu \not\equiv 0$ and u_ν is continuous for all x. Because of the weak convergence

$$0 = \lim_{k \to \infty} \int u_\nu \, d(\mu_k - \mu) = \lim_{k \to \infty} \int (u_{\mu_k} - u_\mu) \, d\nu \geq$$

$$\geq (\rho - r)\nu(G_n) \text{ which implies } \nu(G_n) = \nu(X) = 0.$$

giving us the desired contradiction and the claim is proved.

Since G_n is compact we will even have $C_K^*(G_n) = 0$. It follows that

$$C_K^*(\bigcup_{r,\rho} \bigcup_n G_n) \leq \sum_{r,\rho} \sum_n C_K^*(G_n) = 0.$$

If $\liminf_{n \to \infty} u_{\mu_n}(x) > u_\mu(x)$, then $x \in G_n$ for some r, ρ. We conclude that outside $\bigcup_{r,\rho,n} G_n$, we have $\liminf_{n \to \infty} u_{\mu_n}(x) = u_\mu(x)$. $\square$

LEMMA 5.15. *To an open set $\mathcal{O}$ there exists a measure μ such that*

(i) $u_\mu = 1$ *on $\mathcal{O}$ except on a set of outer capacity 0.*
(ii) $u_\mu \leq 1$ *everywhere.*
(iii) $\mu(X) = C_K(\mathcal{O})$.

PROOF. Take compact sets $F_n \nearrow \mathcal{O}$ and their equilibrium distributions μ_n giving us, by the use of Theorem 5.5, $u_{\mu_n} = 1$ p.p. on $F_n, \mu_n \in \Gamma_{F_n}, \mu_n(F_n) \nearrow C_K(\mathcal{O})$.

The set $F_n \cap \{x : u_{\mu_n}(x) \leq 1 - \frac{1}{k}\}$ is a closed set of capacity 0 and thus of outer capacity 0. The set $F_n \cap \{x : u_{\mu_n}(x) < 1\}$ is a denumerable union of such sets and thus of exterior capacity 0.

Let us now choose a subsequence, also denoted by $\{\mu_n\}$, such that $\mu_n \to \mu$. We know that $u_{\mu_n}(x) = 1$ on F_n except on S_n with $C_K^*(S_n) = 0$. Let $S = \bigcup S_n$, then $C_K^*(S) = 0$. We now use lemma 5.14 and see that $\liminf_{n \to \infty} u_{\mu_n}(x) = u_\mu(x)$, except on T with $C_K^*(T) = 0$. Hence, if $x \notin S \cup T$ but $x \in \mathcal{O}$ we have

$$1 = \liminf u_{\mu_n}(x) = u_\mu(x),$$

which proves (i) because it is clear that $C_K^*(S \cup T) = 0$. The other two parts, (ii) and (iii), are easily obtained,

$$u_\mu(x) \leq \liminf u_{\mu_n}(x) \leq 1 \quad \text{and}$$

$$\mu(X) = \lim \mu_n(X) = C_K(\mathcal{O}).$$

The lemma is proved. $\square$

PROOF OF THEOREM 5.11.. Let $E_n \nearrow E$ and choose open sets $\mathcal{O}_n \supset E_n$ and μ_n with respect to $\mathcal{O}_n$ in the spirit of lemma 5.15 such that $u_{\mu_n}(x) = 1$ on $\mathcal{O}_n$ except for a set of outer capacity 0 and such that

$$\mu_n(X) = C_K(\mathcal{O}_n) \leq C_K^*(E_n) + \frac{1}{n}.$$

Hence we can take a weakly convergent subsequence of $\{\mu_n\}$ such that $\mu_n \to \mu$; as usual we do not change the name of the sequence. Then $u_\mu(x) = \liminf_{n \to \infty} u_{\mu_n}(x) = 1$ for all $x \in E$ except when $x \in S$ where $C_K^*(S) = 0$.

Let $\mathcal{O}_\epsilon := \{u_\mu(x) > 1 - \epsilon\} \supset E \setminus S$. Again, use lemma 5.15 to choose a measure, this time μ_ϵ with respect to $\mathcal{O}_\epsilon$ giving,

$$C_K(\mathcal{O}_\epsilon) = \mu_\epsilon(\mathcal{O}_\epsilon) \leq (1 - \epsilon)^{-1} \int u_\mu \, d\mu_\epsilon =$$

$$= (1 - \epsilon)^{-1} \int u_{\mu_\epsilon} \, d\mu \leq (1 - \epsilon)^{-1} \mu(X).$$

Let us now use this in,

$$C_K^*(E) \leq C_K^*(E \setminus S) + C_K^*(S) \leq C_K^*(\mathcal{O}_\epsilon) \leq$$

$$\leq \mu(X)/(1 - \epsilon) \leq \lim_{n \to \infty} C_K^*(E_n)/(1 - \epsilon).$$

The opposite inequality is trivial, and we have proved Theorem 5.11. $\square$

We are now in a position to prove the remark following Definition 5.2: *If u_ν is a bounded potential and E is an analytic set with $C_K(E) = 0$[5], then $\nu(E) = 0$.*

PROOF. Since all analytic sets are capacitable, we have,

$$C_K(E) = C_K^*(E) = \inf_{\mathcal{O} \supset E} C_K(\mathcal{O}).$$

Let us now fix an open set $\mathcal{O}$ such that, for a given $\epsilon > 0$, $C_K(\mathcal{O}) \leq C_K^*(E) + \epsilon$. Since u_ν is a bounded potential we can, without loss of generality, assume that $u_\nu \leq 1$. If F is a compact subset of $\mathcal{O}$ and $\nu_1 = \nu|_F$, we have $\nu_1 \in \Gamma_F$ and

$$\nu(F) = \nu_1(F) \leq C_K(\mathcal{O})$$

leading to

$$\nu(E) \leq \nu(\mathcal{O}) = \sup_{F \subset \mathcal{O}} \nu(F) \leq C_K(\mathcal{O}) \leq C_K^*(E) + \epsilon = 0 + \epsilon.$$

Hence, $\nu(E) = 0$ and the remark is proved. $\square$

EXERCISE 5.3. *Assume that E is a bounded set such that there exists a measure ν with $u_\nu(x) \geq a$ on E and that $\|\nu\| \leq b$. Prove that $C_K(E) \leq b/a$.*

EXERCISE 5.4. *Discuss what parts of Theorems 4.3 and 4.4 will remain true if we do not assume that the function H is convex.*

EXERCISE 5.5. *Assume that F is a compact set and that $\mathcal{O}$ is a bounded open set. Prove in as simple a way as you can that*

$$C_K(\mathcal{O} \cup F) \leq C_K(\mathcal{O}) + C_K(F).$$

EXERCISE 5.6. *Assume that $\{E_n\}$ is a sequence of capacitable sets. Prove that*

$$C_K(\cup E_n) \leq \sum C_K(E_n).$$

EXERCISE 5.7. *Let F be a compact set and let $\mathcal{O}_n = \{x \in \mathbb{R}^N : d(x, F) < 1/n\}$/ Prove that $\lim_{n \to \infty} C_K(\mathcal{O}_n) = C_K(F)$.*

EXERCISE 5.8. *Prove that there are cases when Problem (B) does not have a unique solution. Discuss the case when the compact set F is the unit circle and $K(r) = \log^+(2/r)$ in the plane. As possible distributions of mass, we can consider normalized Lebesgue measure on the circle and $c\delta_0$, where δ_0 is the Dirac measure at the origin and c is a constant.*

---

[5]or E is a subset of an analytic set of capacity 0.

6. Hausdorff measures and capacities

THEOREM 6.1. *For any bounded set E,*

$$C_K(E) > 0 \Rightarrow \Lambda_{(\bar{K})^{-1}}(E) = \infty$$

where $\bar{K}(r) = r^{-N} \int_0^r K(t) t^{N-1} dt$.

Conversely, if E is analytic and $\Lambda_h(E) > 0$ for a measure function h with $\int_0 K(r) dh(r) < \infty$ then $C_K(E) > 0$.

Example. Let $K(r) = r^{\alpha-N}$ then

$$\bar{K}(r) = r^{-N} \int_0^r t^{\alpha-1} dt = \frac{1}{\alpha} r^{\alpha-N}$$

and $1/\bar{K}(r) = \alpha r^{N-\alpha}$.

Conversely, if $h(r) = \alpha r^{N-\alpha}$ and as above $K(r) = r^{\alpha-N}$ we will have,

$$h(r) \approx \frac{1}{K(r)} \text{ and } \int_0 K(r) dh(r) = \int_0 K(r) d(r^{N-\alpha}) = \infty.$$

From the example above, we see that the two concepts, Hausdorff measure and capacity, are intricately connected.

PROOF OF THEOREM 6.1.. If $C_K(E) > 0$, then there exists a measure $\mu \not\equiv 0$ such that $u_\mu(x) \le 1$ on E and supp$(\mu) \subset E$. Pick a compact subset F_1 of E and its restricted measure $\mu_1 = \mu|_F$. Then $\mu_1 \not\equiv 0$ and u_{μ_1} is a uniformly convergent potential.

Claim. There exists a decreasing function, $K_1(r)$ and a constant M such that $\frac{K(r)}{K_1(r)} \to 0$ as $r \to 0$ and such that $\int K_1(|x-y|) d\mu_1(y) \le M$.

Proof of the claim. To each $\delta > 0$, there exists a $\eta(\delta)$ such that

$$\int_{|x-y|<\eta} K(|x-y|) d\mu_1(y) \le \delta, \ \forall x \in F_1.$$

Let us repeat the argument for a sequence $\{\delta_n\} = \{2^{-n}\}$ and $\eta_n = \eta(2^{-n})$. Then , with the function A, which is to be defined later, we have

$$\int_{|x-y|<a} A(|x-y|) K(|x-y|) d\mu_1(y) \le$$

$$\le \sum_n \int_{\eta_{n+1}<|x-y|<\eta_n<a} A(|x-y|) K(|x-y|) d\mu_1(y) \le$$

$$\le \sum_{\eta_n<a} A(\eta_{n+1}) \int_{|x-y|<\eta_n} K(|x-y|) d\mu_1(y) \le \sum_{\eta_n<a} A(\eta_{n+1}) 2^{-n}.$$

Choose A so that $A(t) \to \infty$ as $t \searrow 0$ and so that

$$\sum_1^\infty A(\eta_{n+1}) 2^{-n} < \infty.$$

Taking $K_1(r) = A(r) K(r)$, we see that the claim is proved.

Next, let us cover F_1 by finitely many balls $\{S_\nu\} = \{\{|x-x_\nu| < r_\nu\}\}$ where the radii are limited, $r_\nu \le \rho$. Then we have,

$$M_1(2r_\nu)^N \ge \int_{|x-x_\nu|\le 2r_\nu} dx \int_{|y-x_\nu|\le r_\nu} K_1(|x-y|) d\mu_1(y) =$$

$$= \int_{|y-x_\nu|\le r_\nu} d\mu_1(y) \int_{|x-x_\nu|\le 2r_\nu} K_1(|x-y|) dx \ge$$

$$\geq \int_{|y-x_\nu| \leq r_\nu} d\mu_1(y) \int_{|x| \leq r_\nu} K_1(x)\, dx \geq \mu_1(S_\nu) C_N \int_0^{r_\nu} K_1(t) t^{N-1}\, dt.$$

Hence,

$$\mu_1(F_1) \leq \sum \mu_1(S_\nu) \leq \sum \frac{CM_1}{K_1(r_\nu)} \leq CM_1 \sup_{r<\rho} \frac{\bar{K}(r)}{K_1(r)} \sum \frac{1}{\bar{K}(r_\nu)}.$$

If the covering has been chosen in such a way that $\sum \frac{1}{\bar{K}(r_\nu)} \approx \Lambda_{\frac{1}{K}}^{(\rho)}(F_1)$, we see that

$$\Lambda_{\frac{1}{K}}^{(\rho)}(F_1) \geq \frac{\mu_1(F_1)}{CM_1 \sup_{r<\rho} \frac{K(r)}{K_1(r)}}$$

and due to the claim this tends to ∞ as $\rho \to 0$, i.e.

$$\Lambda_{\frac{1}{K}}^{(\rho)}(E) \geq \Lambda_{\frac{1}{K}}^{(\rho)}(F_1) = \infty,$$

and the first part of the Theorem is proved. Let us now turn to the second part and assume that E is analytic. We then know that there exists a closed set $F \subset E$ with $M_h(F) > 0$. Furthermore, there is a measure μ on F with

$$\phi(r) := \mu\{x : |x - x_0| < r\} \leq h(r)$$

such that $\mu(F) \geq a M_h(F) > 0$. We got this from Theorem 2.11.

The potential can be integrated by parts, which will be shown later, giving,

$$u_\mu(x_0) = \int K(r)\, d\phi(r) = -\int \phi(r)\, dK(r) \leq$$

$$\leq -\int h(r)\, dK(r) = \int K(r)\, dh(r) = \text{Const.} < \infty.$$

Hence, $C_K(E) > 0$.

The integration by parts is allowed because we have,

$$K(\rho)h(\rho) = \int_0^\rho K(\rho)\, dh(r) \leq \int_0^\rho K(r)\, dh(r) \to 0, \quad \text{as } \rho \to 0.$$

The ϕ-integral can be treated in a similar way and the proof of Theorem 6.1 is finished. □

6.1. Coverings. Let us first present, without a proof, a useful covering lemma which is due to Besicovitch.

LEMMA 6.2. *Suppose that $E \subset \mathbb{R}^N$ is covered by balls such that each $x \in E$ is the center of a ball $S(x)$ of radius $r(x)$. If $\sup r(x) < \infty$, then it is possible to select a countable number of balls $\{S(x_k)\}$ from $\{S(x)\}_{x \in E}$ such that $E \subset \bigcup S(x_k)$ and such that each $x \in \mathbb{R}^N$ is covered by at most $B(N)$ balls, where $B(N)$ is a constant depending only on the dimension N.*

For the proof, see [36, Theorem 1.3.5] or [29, Lemma 4.3.2].

Remark. The condition $\sup r(x) < \infty$, is not needed if we assume E to be bounded.

Let us now assume that E is a bounded set, let $A(r)$ be the minimal number of balls, with radius r, needed to cover E and let $A_{CL}(r)$ be the number of balls from the covering lemma 6.2. Let $\mathcal{B}$ be a minimal covering and $\mathcal{B}_{CL}$ a covering from the covering lemma.

Consider $B \in \mathcal{B}$ and $B' \in \mathcal{B}_{CL}$ such that $B \cap B' \neq \emptyset$. The sum of the volumes of all such balls B' is less than $B(N)(3r)^N C_N$, where RNC_N is the volume of a ball with radius r. The number of such balls B' can be estimated,

$$\sharp(B') \leq \frac{B(N)3^N RNC_N}{r^N C_N} = B(N)3^N.$$

The total number of balls in $\mathcal{B}_{CL}$ is $A_{CL}(r) \leq A(r)B(N)3^N$ and we conclude,

$$A(r) \leq A_{CL}(r) \leq A(r)B(N)3^N.$$

If we do not need too many balls to cover a set then the capacity of the set must be small. The following Theorem gives a concrete relation.

THEOREM 6.3. *Assume that the set E can be covered by $A(r)$ closed balls of radii $\leq r$ and that $-\int_0 \frac{dK(r)}{A(r)} = \infty$, then $C_K(E) = 0$.*

PROOF. WLOG, assume all balls have radius r. Choose $A(r)$ as the minimal number of such balls needed. Assume $C_K(E) > 0$ then the equilibrium measure μ exists. We also assume that E is analytic. Since the capacity is positive the energy integral is bounded, i.e. $I(\mu) < \infty$. Define $\mu(r,a) := \mu(\{x : |x-a| \leq r\})$. The plan is to get a contradiction by showing that the energy integral is unbounded.

$$I(\mu) = \int d\mu(y) \int K(|x-y|)\, d\mu(x) = \int d\mu(y) \int_0^\infty K(r)\, d\mu(r,y).$$

Since the potential is bounded we can integrate by parts,

$$\int d\mu(y) \int_0^\infty K(r)\, d\mu(r,y) = \int d\mu(y) \int_0^\infty (-\mu(r,y))\, dK(r)\mu(r,y) \geq$$

$$\geq \sum_{n=0}^\infty \int_{2^{-n-1}}^{2^{-n}} (-dK(r)) \int \mu(2^{-n-1}, y)\, d\mu(y).$$

Assume that we have a covering $E \subset \bigcup_{\nu=1}^{A_n} S_\nu^{(n)}$ where $S_\nu^{(n)}$ are closed balls with radius 2^{-n} and that $A_n = A(2^{-n})$ is the minimal number of balls needed. There is a constant, C, such that

$$\sum_{n=0}^\infty \int_{2^{-n-1}}^{2^{-n}} (-dK(r)) \int \mu(2^{-n-1}, y)\, d\mu(y) \geq$$

$$\geq \sum_{n=0}^\infty \int_{2^{-n-1}}^{2^{-n}} (-K'(r)dr) \frac{1}{C} \sum_{\nu=0}^{A_{n+2}} \int_{S_\mu^{(n+2)}} \mu(2^{-n-1}, y)\, d\mu(y) \geq$$

$$\geq \frac{1}{C} \sum_{n=0}^\infty \int_{2^{-n-1}}^{2^{-n}} (-K'(r)dr) \sum_{\nu=0}^{A_{n+2}} \mu(S_\mu^{(n+2)})^2.$$

Since,

$$\mu(E)^2 \leq \left(\sum_{\nu=0}^{A_n} \mu(S_\mu^{(n)}) \right)^2 \leq A_n \sum_{\nu=0}^{A_n} \mu(S_\mu^{(n)})^2$$

we get,

$$\frac{1}{C} \sum_{n=0}^\infty \int_{2^{-n-1}}^{2^{-n}} (-K'(r)dr) \sum_{\nu=0}^{A_{n+2}} \mu(S_\mu^{(n+2)})^2 \geq$$

$$\geq \frac{1}{C} \sum_{n=0}^\infty \frac{\mu(E)^2}{A_{n+2}} (K(2^{-n-1}) - K(2^{-n})).$$

If we now cover balls with radius 2^{-n} by balls of radius 2^{-n-2} we get the relation: $A_{n+2} \leq C_N A_n$ where C_N is a constant only depending on the dimension N. Thus,

$$\frac{1}{C} \sum_{n=0}^{\infty} \frac{\mu(E)^2}{A_{n+2}} (K(2^{-n-1}) - K(2^{-n})) \geq$$

$$\geq \frac{1}{CC_N} \sum_{n=0}^{\infty} \frac{\mu(E)^2}{A_n} (K(2^{-n-1}) - K(2^{-n})).$$

Retracking the estimates,

$$I(\mu) \geq \frac{1}{CC_N} \sum_{n=0}^{\infty} \frac{\mu(E)^2}{A_n} (K(2^{-n-1}) - K(2^{-n})).$$

The condition: $-\int_0 \frac{dK(r)}{A(r)} = \infty$ in the Theorem gives us that

$$\sum \frac{(K(2^{-n-1}) - K(2^{-n}))}{A_n} = \infty.$$

It follows that $I(\mu) \geq \infty$ which is the desired contradiction. $\square$

6.2. Cantor sets. Let us now study some examples of sets which catch some of the subtle interaction between Hausdorff measures and capacities. It turns out that the generalized Cantor sets are suitable for this task.

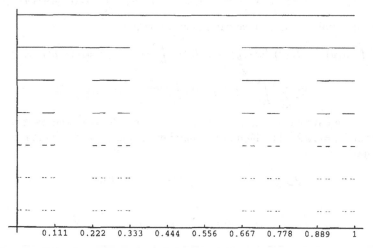

| 0.111 | 0.222 | 0.333 | 0.444 | 0.556 | 0.667 | 0.778 | 0.889 | 1 |

FIGURE 6.1. A construction of the usual Cantor set in one dimension , i.e. $\{l_n\} = \{\frac{1}{3^n}\}$. The generalization to higher dimensions is straightforward.

THEOREM 6.4. *Let E be a N-dimensional Cantor set constructed from the sequence $\{l_n\}_1^{\infty}$, then E has positive capacity if and only if the series $\sum_{n=1}^{\infty} 2^{-nN} K(l_n)$ is convergent.*

PROOF. Assume that $\sum_{n=1}^{\infty} 2^{-nN} K(l_n) < \infty$ and let E_n be the n:th stage of the construction. We have then intervals of length l_n. We choose measures μ_n with support in E_n such that $\mu_n(E_n) = 1$ and such that μ_n is uniformly distributed on all intervals. The density will then be $\frac{1}{2^{nN} l_n^N}$.

There are at most $(2 \cdot 2 + 1)^N$ intervals from E_n within distance l_{n-1} from $x_0 \in E_n$. Repeating this argument, we will have, $(2 \cdot 2^\nu + 1)^N$ intervals from E_n within distance $l_{n-\nu}$ from $x_0 \in E_n$.

Let us first prove $\sum_{n=1}^\infty 2^{-nN} K(l_n) < \infty \Rightarrow C_K(E) > 0$. Let us estimate the potentials,

$$\int K(x_0 - y) \, d\mu_n(y) \le C[l_n^{-N} \int_{|t|<l_n} 2^{-nN} K(t) \, dt +$$

$$+ \sum_{\nu=0}^{n-1} \int_{l_{\nu+1} \le |x_0-y| \le l_\nu} K(x_0 - y) \, d\mu_n(y)] = C(I + \sum J_\nu),$$

where the constant C only depends on the dimension and as usual we do not change notation for different constants like that. We now use the above result that the number of intervals in E_n within a distance l_ν is less than $(2 \cdot 2^{n-\nu} + 1)^N$,

$$J_\nu \le K(l_{\nu+1}) \int_{|x_0-y| \le l_\nu} d\mu_n(y) \le K(l_{\nu+1}) \frac{(2 \cdot 2^{n-\nu} + 1)^N l_n^N}{2^{nN} l_n^N} \le$$

$$\le C K(l_{\nu+1}) 2^{-\nu N}.$$

The first part can also be estimated,

$$I \le (l_n 2^n)^{-N} \sum_{\nu=n}^\infty \int_{l_{\nu+1} \le |t| \le l_\nu} K(t) \, dt \le$$

$$\le C[K(l_{n+1}) 2^{-nN} + K(l_{n+2}) \left(\frac{l_{n+1}}{l_n}\right)^N 2^{-nN} + \cdots \le C \sum_{\nu=n}^\infty K(l_\nu) 2^{-\nu N},$$

since $l_{n+1} \le \frac{1}{2} l_n$ which gives,

$$\left(\frac{l_{n+1}}{l_n}\right)^N 2^{-nN} \le 2^{-(n+1)N}.$$

We immediately have,

$$C \sum_{\nu=n}^\infty K(l_\nu) 2^{-\nu N} \to 0, \text{ as } n \to \infty.$$

The remainder term will then be,

$$\sum_1^{n-1} J_\nu \le C \sum_1^{n-1} K(l_{\nu+1}) 2^{-(\nu+1)N} < C \sum_1^\infty 2^{-mN} K(l_m).$$

If we now let $\mu_n \longrightarrow \mu$, we will get,

$$u_\mu(x_0) = \int K(x_0 - y) \, d\mu(y) \le C \sum_1^\infty 2^{-mN} K(l_m) < \infty$$

by the assumption. This holds for every $x_0 \in F$ and $\mu(E) = 1$ by the way it was constructed, why we conclude $C_K(E) > 0$. The first part of the proof is finished.

Conversely, we prove that $\sum_{n=1}^\infty 2^{-nN} K(l_n) = \infty \Rightarrow C_K(E) = 0$.

Let, as above, $A(r)$ be the minimal number of balls with radius r such that their union covers E. Then we have $A(l_n) \le C 2^{nN}$. Applying Theorem 6.3 we deduce that

$$- \int_0 \frac{dK(r)}{A(r)} = \sum^\infty - \int_{l_{n+1}}^{l_n} \frac{dK(r)}{A(r)} \ge$$

$$\ge \sum^\infty \frac{K(l_{n+1}) - K(l_n)}{A(l_{n+1})} \ge C \sum^\infty \frac{K(l_{n+1}) - K(l_n)}{2^{nN}} =$$

$$= C \sum_{}^{\infty} \frac{K(l_{n+1})}{2^{nN}} - C \sum_{}^{\infty} \frac{K(l_n)}{2^{nN}} = C \sum_{}^{\infty} \frac{K(l_n)}{2^{nN}} (2-1) + C_1 = \infty.$$

The theorem is proved. □

6.3. A Cantor type construction. We will now show that the first condition of Theorem 6.1 is best possible. We will for simplicity restrict ourselves to the linear case (i.e. $N = 1$). At stage r, which means that the length of the intervals are r, we have $n(r)$ intervals and the proof requires:

(i) $rn(r) \to 0$ as $r \to 0$.

(ii) $n(r)h(r) \to 0$ as $r \to 0$.

(iii) $\frac{\bar{K}(r)}{n(r)} \to 0$ as $r \to 0$ (for the definition of $\bar{K}(r)$, we refer to Theorem 6.1).

We note that, (ii) implies that the Hausdorff measure $\Lambda_h(E) = 0$ and (i) means that the Lesbegue measure of the union of the intervals at stage r is rapidly tending to zero. The last requirement is just a technical point needed in the proof of the following theorem.

THEOREM 6.5 (A). *Let* $N = 1$. *With* K, h *satisfying* $h(r)\bar{K}(r) \to 0$, *there is a Cantor set* E *with the properties* $C_K(E) > 0$ *and* $M_h(E) = 0$.

PROOF. Since we have assumed integrability for the kernel we have, $\int_0 K(r)dr < \infty$, which implies,

(10) $\eta(r) := r\bar{K} \to 0$ as $r \to 0$

We will beside η also define $\epsilon(r) := h(r)\bar{K}(r)$ which by assumption also tends to zero. Next we find $\eta_1(r)$ tending to zero, when $r \to 0$, so slowly that $\frac{\epsilon(r)}{\eta_1(r)} \to 0$ and $\frac{\eta(r)}{\eta_1(r)} \to 0$. One way to construct such function is , $\eta_1 := [\max(\epsilon(r), \eta(r))]^{\frac{1}{2}}$. Furthermore, define $n(r) := [\frac{\bar{K}(r)}{\eta_1(r)}]$, where $[t]$ is the integer part of t. Let us check the above requirements

(i) $rn(r) \approx \frac{r\bar{K}(r)}{\eta_1(r)} = \frac{\eta(r)}{\eta_1(r)} \to 0$ as $r \to 0$,

(ii) $n(r)h(r) \approx \frac{h(r)\bar{K}(r)}{\eta_1(r)} = \frac{\epsilon(r)}{\eta_1(r)} \to 0$ as $r \to 0$,

(iii) $\frac{\bar{K}(r)}{n(r)} \approx \eta_1(r) \to 0$ as $r \to 0$.

The requirements are fulfilled and we are now ready to construct the Cantor set. We start with the closed interval $[0, 1]$ and throw away $n_1 - 1$ open intervals of length s_1 symmetrically distributed on $(0, 1)$. There will then be n_1 intervals of length r_1 left. The connection will then be $n_1 r_1 + (n_1 - 1)s_1 = 1$. Furthermore, we will assume $n_1 r_1 \ll 1$.

In the next step we will repeat the above argument for the interval $[0, 1]$ changed to the n_1 intervals of length r_1. Giving, n_2 intervals of length r_2 and $n_2 r_2 + (n_2 - 1)s_2 = r_1$. We will here assume $n_2 r_2 \ll r_1$.

We now proceed inductively. If we have found the sequences $\{n_j\}_{j=1}^{k-1}$ and $\{r_j\}_{j=1}^{k-1}$, we choose n_k so large that if r_k is defied by the relation

$$\prod_{j=1}^{k} n_j \approx \frac{\bar{K}(r_k)}{\eta_1(r_k)} \approx n(r_k),$$

we have $n_k r_k < r_k n(r_k) \ll r_{k-1}$.

It is now possible to show, see [7, page 35], that the potential is not disturbed too much by changes of the mass distribution "far away". In the limit the potential will still be bounded which in turn implies that $C_K(E) > 0$. On the other hand (ii) tells us that $\Lambda_h(E) = 0$. Theorem 6.5 is proved. □

Carleson also gives a theorem on the opposite situation.

THEOREM 6.6. *There exists a set E of the type constructed above such that $M_h(E) = 0$ for all h such that $\int_0 \frac{h(r)}{r} dr < \infty$, while $C_K(E) > 0$ with $K(r) = \log \frac{1}{r}$.*

PROOF. See [7, Theorem 5, p. 35]. □

Remark of V. Eiderman.

"It follows from Theorem 6.5 (A) that Theorem 6.1 is sharp. We note, that Theorem 6.5 below is also sharp: it is impossible to replace the condition $h(r)\bar{K}(r) \to 0$ by $h(r)K(r) \to 0$.

Claim. *Let $N = 1$, then there exists a measure function h and a kernel K such that*

(1) $h(r)K(r) \to 0$, $r \to 0$,
(2) $\int_0 K(r) dr < \infty$,

and such that for any bounded set E,

$$C_K(E) > 0 \Rightarrow M_h(E) > 0.$$

Proof of the claim. Let $k(r) = \frac{1}{r}\log^{-k}\frac{1}{r}$, $k > 1$. Then $\bar{K}(r) = \frac{1}{k-1}\frac{1}{r}\log^{1-k}\frac{1}{r}$, and for $h(r) := 1/\bar{K}(r)$ we have $h(r)K(r) \to 0$ as $r \to 0$. The claim follows now from Theorem 6.1.

In the Russian translation of [7], V. Havin has made several remarks on the proof of Carleson. The original statement of Carleson and the beginning of the proof according to Havin are as follows.

THEOREM 6.5. *For any kernel K and any measure function h such that*

$$\liminf_{r \to 0} h(r)\bar{K}(r) = 0,$$

there is a set E such that $C_K(E) > 0$ and $M_h(E) = 0$.

PROOF. Define $\{r_\nu\}$ tending to zero so that

(11) $$h(r_\nu)\bar{K}(r_\nu) = \epsilon_\nu^2 \to 0.$$

We shall prove that there is a subsequence $\{r_{\nu_i}\}$ such that

(12) $$n_{i+1}r_{\nu_{i+1}} < n_i r_{\nu_i}$$

where

$$n_1 = 1, \ n_i = n_{i-1}\left[\bar{K}(r_{\nu_i})/(\epsilon_{\nu_i}n_{i-1})\right], \ i > 1$$

(there is a slight improvement of the argument in [7, Russian translation p.41].) Without loss of generality, we assume that

$$\liminf_{r \to 0} h(r)/r > 0$$

(in the opposite case, $M_h([0,1]) = 0$). Hence

$$n_{i+1}r_{\nu_{i+1}} \leq \bar{K}(r_{\nu_{i+1}})c_{\nu_{i+1}}^{-1}r_{\nu_{i+1}} \leq C\bar{K}(r_{\nu_{i+1}})h(r_{\nu_{i+1}})/\epsilon_{\nu_{i+1}} = C\epsilon_{\nu_{i+1}}.$$

Thus, if n_i and r_{ν_i} are given, then (12) will hold if ν_{i+1} is large enough. From (11), we deduce that $M_h(E) = 0$.

We can now continue the proof as on line 5 from below p. 34 in [7]. □

Finally, we would like to mention that Havin's Remark 17, p. 112 in [7, Russian translation] is not correct in general but only under additional assumptions on h (in Havin's reference [27], it is assumed that h is convex). For details, we refer to Eiderman [12]".

Let us end this section by recollecting some result about the connections between capacities and Hausdorff measures.

- If $h = 1/\bar{K}$ then $C_K(E) > 0 \Rightarrow \Lambda_h(E) = \infty$.
- If $h(r)\bar{K}(r) \to 0$ as $r \to 0$, then there is a set E such that $C_K(E) > 0$ but $\Lambda_h(E) = 0$.
- If E analytic, $\Lambda_h(E) > 0$ and $\int_0 K(r)\, dh(r) < \infty$, then $C_K(E) > 0$. The opposite is not true.
- Theorem 6.6 gives the existence of a set E such that $C_K(E) > 0$ but $M_h(E) = 0$ for all h as in Theorem 6.6.
- The condition $\int_0 K(r)\, dh(r) < \infty$ is "almost necessary" for the implication $\Lambda_h(E) > 0 \Rightarrow C_K(E) > 0$, see [12]".

7. Two Extremal Problems

Let $F \subset \mathbb{R}^N$ be compact.

DEFINITION 7.1 (CHEBYCHEV'S CONSTANT).

$$M_n := M_n(F) = \sup_{\{x_\nu\}_1^n} \inf_{x \in F} \frac{1}{n} \sum_1^n K(x - x_\nu).$$

DEFINITION 7.2 (GENERALIZED DIAMETER).

$$D_n := D_n(F) = \inf_{\{x_\nu \in F\}_1^n} \binom{n}{2}^{-1} \sum_{1 \le \nu < j \le n} K(x_\nu - x_j).$$

THEOREM 7.3. If $C_K(F) > 0$ and F is compact, then the limits $\lim_{n \to \infty} M_n = M$ and $\lim_{n \to \infty} D_n = D$ exist and

$$M^{-1} = D^{-1} = C_K(F).$$

If $C_K(F) = 0$, then $M_n \to \infty$ and $D_n \to \infty$.

PROOF. We will prove a number of claims.

(1) $D_{n+1} \le M_n$

Proof of the claim By continuity, find $\{x_\nu\}_1^{n+1}$ such that

$$D_{n+1} = \frac{2}{n(n+1)} \sum_{i<j} K(x_i - x_j) = \frac{1}{n(n+1)} \sum_{i \ne j} \sum K(x_i - x_j).$$

Let now ν be given. Define

$$x_i' = \begin{cases} x_i & \text{if } i \ne \nu \\ \text{arbitrary} & \text{if } i = \nu. \end{cases}$$

The minimality then gives us,

$$\sum_{i \ne j} \sum K(x_i - x_j) \le \sum_{i \ne j} \sum K(x_i' - x_j').$$

All terms with $i, j \ne \nu$ will be equal and can therefore be canceled.

$$\sum_{i \ne \nu} K(x_i - x_\nu) + \sum_{j \ne \nu} K(x_j - x_\nu) \le \sum_{i \ne \nu} K(x_i' - x_\nu') + \sum_{j \ne \nu} K(x_j' - x_\nu').$$

If we define $A_\nu(x) := \sum_{j\neq\nu} K(x - x_j)$ we will have

$$\min A_\nu(x) = A_\nu(x_\nu) =$$

$$= \inf\left((\sum_{j=1}^{n+1} K(x - x_j)) - K(x - x_\nu) \right) \leq nM_n.$$

Hence,

$$D_{n+1} = \frac{1}{n(n+1)} \sum_{i=1}^{n+1} \sum_{i\neq j} K(x_i - x_j) \leq$$

$$\leq \frac{1}{n(n+1)} \sum_{i=1}^{n+1} A_i(x_i) \leq \frac{1}{n(n+1)} nM_n(n+1) = M_n.$$

The first claim is proved.

(2) $M_n \leq V = \frac{1}{C_K(F)}$.

There is an equilibrium distribution μ on F such that $\mu(F) = 1$ and $\int K(x - y)\,d\mu(y) \leq V$ everywhere (due to the strong maximum principle , Theorem 4.3).

$$\inf_{x\in F} \frac{1}{n} \sum_1^n K(x - x_\nu) \leq \int \frac{1}{n} \sum_1^n K(x - x_\nu)\,d\mu(x) \leq V.$$

This implies $M_n \leq V$.

(3) $V \leq \lim_{n\to\infty} D_n$.

We define the truncated kernel, $K_A(r) := \min(K(r), A)$ and choose $\{x_\nu\}$ as a minimizing sequence for D_n. Let μ_n be the measure with mass $\frac{1}{n}$ at each x_ν. The energy integral will then be,

$$I_{K_A}(\mu_n) = \frac{1}{n} \int \sum_1^n K_A(y - x_\nu)\,d\mu_n(y) \leq$$

$$\leq \frac{1}{n^2} \sum \sum_{i\neq\nu} K(x_i - x_\nu) + \frac{1}{n} An \frac{1}{n}.$$

If we now take a weakly convergent subsequence $\{\mu_{n_k}\}$ of $\{\mu_n\}$ such that $\mu_{n_k} \rightharpoonup \mu$ and such that $D_{n_k} \to \liminf D_n$, then we have

$$I_{K_A}(\mu) \leq \liminf D_n.$$

Letting $A \to \infty$ we have, since $\operatorname{supp}\mu \subset F$ and $\mu(F) = 1$, $V \leq I(\mu)$. By using the two first claims we put this all together now,

$$V \leq I(\mu) \leq \liminf D_n \leq \limsup D_n \leq \limsup M_n \leq V.$$

So there has to be equality everywhere. Especially, we have $\liminf D_n = \limsup D_n$ and

$$V = I(\mu) = \lim D_n \leq \liminf M_n \leq \limsup M_n \leq V,$$

giving us the existence of the limits and the equality if $C_K(F) > 0$.

If, on the other hand, $C_K(F) = 0$, find a compact $F' \supset F$ such that $C_K(F') > 0$. It is clear that

$$D_n(F) \geq D_n(F') = \frac{1}{C_K(F')} \to \infty \text{ as } F' \searrow F.$$

The proof of Theorem 7.3 is finished. $\square$

7.1. The Classical Case. Let us now after this discussion of the generalized entities M_n and D_n go back to the classical case. This takes place in the plane, i.e. $N = 2$. As usual, F is a compact set. We define,

DEFINITION 7.4 (DIAMETER OF ORDER n).

$$d_n(F) = \max_{z_k \in F} \prod_{1 \leq i < j \leq n} |z_i - z_j|^{\frac{2}{n(n-1)}}.$$

Ahlfors shows in [1] that $d_n \leq d_{n-1}$ and we can define,

DEFINITION 7.5 (TRANSFINITE DIAMETER OF F).

$$d_\infty(F) = \lim d_n(F).$$

Let us now consider polynomials, $P_n(z) = z^n + a_1 z^{n-1} + \cdots + a_n$ and define,

DEFINITION 7.6 (CHEBYCHEV POLYNOMIAL OF ORDER n).

$$\rho_n^n(F) = \min_{P_n} \max_{z \in F} |P_n(z)|.$$

The two newly defined functions are related by the following Theorem.

THEOREM 7.7.

$$\lim_{n \to \infty} \rho_n = d_\infty.$$

PROOF. See [1]. □

This theorem is related to the theorems above as we can see from this analysis.

$$\log \frac{1}{d_n} = \inf \frac{2}{n(n-1)} \sum_{i<j} \log \frac{1}{|z_i - z_j|}.$$

Let us choose the classical kernel in the plane, i.e. $K(r) = \log(1/r)$. Assume furthermore that $|z_j| \subset \{|z| < \frac{1}{2}\}$ and that $F \subset \{|z| < \frac{1}{2}\}$ which will give us $K(|z_i - z_j|) = \log^+ \frac{1}{|z_i - z_j|}$ so that we will not have any problem with the change of sign of the log–function.

It is always possible to write polynomials $P_n(z)$ in the form $P_n(z) = \prod_1^n (z - z_\nu)$. Hence,

$$\log \frac{1}{\rho_n} = \max_{z \in F} \min \frac{1}{n} \sum_1^n \log\left(\frac{1}{|z - z_\nu|}\right)$$

and we see that in this special case, Theorems 7.3 and 7.7 are equivalent.

We have covered the first four chapters in Carleson's book [7]. Let us quote the titles of the remaining chapters where there are applications of the theory.

  V.: Existence of boundary values.
  VI.: Existence of certain holomorphic functions.
  VII.: Removable singularities for harmonic functions.

For problems discussed in chapter VII, there are further developments in [21] and [31].

8. M. Riesz kernels

We now turn to another way of introducing capacity and follow the presentation in Landkof [29]. In many cases, we can use results from the previous sections of these notes.

Let $N \geq 2$ and $x \in \mathbb{R}^N$. Define the kernel,

$$k_\alpha(x) := A(N, \alpha)|x|^{\alpha - N}, \quad \text{for } 0 < \alpha < N.$$

The kernel that belongs to the Laplace operator in $\mathbb{R}^N$ is $k_2(x)$, i.e. $\Delta k_2 = -C_N \delta$.

The constant in the definition $A(N, \alpha)$ is chosen so that

(13) $$k_\alpha * k_\beta = k_{\alpha + \beta}.$$

That is,

$$A(N, \alpha) = \frac{\pi^{\alpha - \frac{N}{2}} \Gamma\left(\frac{N - \alpha}{2}\right)}{\Gamma\left(\frac{\alpha}{2}\right)},$$

where Γ is the usual gamma–function,

$$\Gamma(z) = \int_0^\infty e^{-t} t^{z-1} \, dt.$$

The natural question arises: "When is $k_\alpha * k_\beta$ defined?" There are obvious problems at infinity. Assume $|x| \ll R$, where R is a constant and $x \neq 0$. We have then,

$$\int_{|y| > R} \frac{dy}{|y|^{N - \alpha}|x - y|^{N - \beta}} \approx \int_{|y| > R} \frac{dy}{|y|^{2N - \alpha - \beta}} \approx$$

$$\approx \int_R^\infty r^{N - 1 - 2N + \alpha + \beta} \, dr = \int_R^\infty r^{\alpha + \beta - N - 1} \, dr$$

which is convergent if and only if $\alpha + \beta < N$. This can be generalized (see [29] for details) by using analytic continuation to: If $\Re(\alpha + \beta) < N$ and $\alpha, \beta \neq N + 2n, n \in \mathbb{N}$, then formula 13 is valid. We will also have, in a distributional sense,

$$k_\alpha * k_{-\alpha} = \delta.$$

Let us now define,

DEFINITION 8.1 (THE LOGARITHMIC KERNEL).

$$k_N = \omega_N \log(1/|x|).$$

Let us assume that $f \in C_0(\mathbb{R}^N)$ and $\int f \, dx = 0$, then

$$\lim_{\alpha \to N} \int k_\alpha(x) f(x) \, dx = \omega_N \int \log \frac{1}{|x|} f(x) \, dx.$$

It follows that $k_N - k_{N/2} * k_{N/2}$ is constant telling us that we have to be careful and think in terms of distributions.

8.1. Potentials. From now on, if nothing else is said, assume all measures to have compact support. We define,

DEFINITION 8.2 (THE POTENTIAL).

$$U_\alpha^\nu(x) = \int k_\alpha(x-y)\,d\nu(y).$$

We did some work in the previous sections which we will take advantage of now by presenting some theorems and use the above results in the proofs.

THEOREM 8.3.   If $2 < \alpha < N$: then $U_\alpha^\mu(x)$ is superharmonic in $\mathbb{R}^N$.
    If $\alpha = 2$: then $U_2^\mu(x)$ is superharmonic in $\mathbb{R}^N$ and $U_2^\mu(x)$ is harmonic outside
    supp μ.
    If $0 < \alpha < 2$: then $U_\alpha^\mu(x)$ is subharmonic outside supp μ.

PROOF. We have

$$\Delta k_\alpha(x) = A(N,\alpha)\frac{(N-\alpha)(2-\alpha)}{|x|^{N-\alpha+2}},$$

where we assume $x \neq 0$. This will give us.

    If $2 < \alpha < N$ **then** : $\Delta k_\alpha(x) < 0$.
    If $\alpha = 2$ **then** : $\Delta k_\alpha(x) = 0$.
    If $0 < \alpha < 2$ **then** : $\Delta k_\alpha(x) > 0$.

Let us first assume $2 \leq \alpha < N$.

We will take as a fact that k_α is superharmonic in $\mathbb{R}^N$. Then we know that $k_{\alpha,n} := \min(k_\alpha, n)$ is superharmonic. Hence $U_n(x) := \int k_{\alpha,n}(x-y)\,d\mu(y)$ is superharmonic because it is continuous and has the superharmonic mean value property which we see from the following estimate (where $d\sigma$ is the normalized surface measure):

$$\int_{|\omega|=1} U_n(x+\rho\omega)\,d\sigma = \iint_{|\omega|=1} k_{\alpha,n}(x+\rho\omega-y)\,d\sigma\,d\mu(y) \leq$$

$$\leq \int k_{\alpha,n}(x-y)\,d\mu(y) = U_n(x).$$

Now, since we have $U_n \nearrow U$ we have that U is semicontinuous from below and that

$$U_n(x) \geq \int_{|\omega|=1} U_n(x+\rho\omega)\,d\sigma \nearrow \int_{|\omega|=1} U(x+\rho\omega)\,d\sigma,$$

where we used monotone convergence. Thus we have,

$$U_n(x) \to U(x) \geq \int_{|\omega|=1} U(x+\rho\omega)\,d\sigma,$$

and we have proved the first statement.

For the second statement we note that it is allowed to differentiate under the integral outside the support of μ.

The third statement is proved in a similar way. □

THEOREM 8.4. If $U_\alpha^\mu(x) \leq M$ on the support of μ, then $U_\alpha^\mu(x) \leq 2^{N-\alpha}M$, $\forall x \in \mathbb{R}^N$.

PROOF. Let $x \notin S_\mu$, the support of μ and let $x' \in S_\mu$ be the nearest point to x. Assume furthermore that $y \in S_\mu$. We have then,

$$|y - x'| \leq |y - x| + |x - x'| \leq 2|y - x|.$$

Hence,

$$\int \frac{A(N, \alpha)\, d\mu(y)}{|x - y|^{N-\alpha}} \leq \int \frac{A(N, \alpha)\, d\mu(y)}{|x' - y|^{N-\alpha}} 2^{N-\alpha},$$

and we obtain

$$U_\alpha^\mu(x) \leq 2^{N-\alpha} M.$$

$\square$

THEOREM 8.5. *Suppose that* $0 < \alpha \leq 2$. *If* $U_\alpha^\mu(x) \leq M$ *p.p. on* $\operatorname{supp}\mu$, *then* $U_\alpha^\mu(x) \leq M \ \forall x \in \mathbb{R}^N$.

PROOF. This has already been proven in Theorem 4.3 (cf. also the remark preceding Theorem 5.8). $\square$

THEOREM 8.6. *Assume that* μ *is a measure in the plane with compact support and define* $U_2^\mu(x) := \int \log \frac{1}{|x-y|}\, d\mu$. *If* $U_2^\mu(x) \leq M$ *p.p. on* $\operatorname{supp} \mu$, *then* $\tilde{U}_2^\mu(x) \leq M \ \forall x \in \mathbb{R}^2$.

PROOF. Define a new measure, ν, by $d\nu(y) := d\mu(ay)$. This will give us

$$\operatorname{supp}\mu \subset \{|x| \leq R\} \Rightarrow \operatorname{supp}\nu \subset \{|x| \leq R/a\}.$$

Choose $a = 2R$, hence, $\operatorname{supp}\nu \subset \{|y| \leq 1/2\}$. Therefore,

$$U^\nu(x) = \int \log \frac{1}{|x - y|} d\mu(ay) = U^\mu(ax) + \log(a)\mu(\mathbb{R}^N).$$

We have also $U^\mu(x) \leq M$ on $\operatorname{supp}\mu$, which is equivalent to

$$U^\nu(x) \leq M + \log(a)\mu(\mathbb{R}^N) = M' \quad \text{on } \operatorname{supp}\nu.$$

The auxiliary function

$$H(t) = \begin{cases} t & \text{if } t \geq 0 \\ 0 & \text{if } t < 0. \end{cases}$$

is convex. Thus we can define a potential which satisfies the strong maximum principle

$$U(x) := \int H(\log \frac{1}{|x - y|}) d\nu(y) = \int \log^+ \frac{1}{|x - y|} d\nu(y).$$

Hence we have, $U(x) = U^\nu(x)$ on the support of ν and $U^\nu(x) \leq U(x) \leq M'$ everywhere, implying

$$U^\mu(x) \leq M \quad \text{everywhere}.$$

The proof is complete. $\square$

THEOREM 8.7. *If* ν *is a signed measure and* $0 < \alpha < N$, *then the energy integral is defined by*

$$I_\alpha(\nu) := \int_{\mathbb{R}^N} U_\alpha^\nu \, d\nu(x),$$

and we have

$$I_\alpha(\nu) \geq 0 \quad \text{and} \quad I_\alpha(\nu) = 0 \Leftrightarrow \nu \equiv 0.$$

Remark on the definition of the integral. Let $\nu = \nu^+ - \nu^-$ giving,

$$\int U_\alpha^\nu \, d\nu = \int (U_\alpha^{\nu^+} - U_\alpha^{\nu^-}) \, d(\nu^+ - \nu^-) =$$

$$= \int U_\alpha^{\nu^+} \, d\nu^+ + \int U_\alpha^{\nu^-} \, d\nu^- - \int U_\alpha^{\nu^-} \, d\nu^+ - \int U_\alpha^{\nu^+} \, d\nu^-.$$

Now, to make this well defined, we consider only measures ν such that $\int U_\alpha^{\nu^+} \, d\nu^- < \infty$ and $\int U_\alpha^{\nu^-} \, d\nu^+ < \infty$.

Proof of the theorem. We will give two different proofs.

PROOF 1.. The case $0 < \alpha \leq 2$ has already been taken care of, see Lemma 5.10. In the case of Riesz kernels we have the Fourier transform,

$$\hat{k}_\alpha(\xi) = C(N, \alpha)|\xi|^{-\alpha}, \quad \text{where } C(N, \alpha) > 0.$$

The previous proof will also work in this case when $2 < \alpha < N$. $\square$

PROOF 2.. We will here use the identity, $k_\alpha = k_{\frac{\alpha}{2}} * k_{\frac{\alpha}{2}}$. Let us insert this into the energy integral,

$$I_\alpha(\nu) = \iint d\nu(x) \, d\nu(y) \int k_{\frac{\alpha}{2}}(x - z) k_{\frac{\alpha}{2}}(z - y) \, dz = \int_{\mathbb{R}^N} dz \left(U_{\frac{\alpha}{2}}^\nu(z) \right)^2 \geq 0.$$

What about the equality, $I_\alpha = 0$? If the energy integral is zero, does it follow that $U_{\frac{\alpha}{2}}^\nu(z) = 0$ a.e.? Let us take a test function in the Schwartzian space, $\varphi \in S$. Then we have,

$$\int (k_{\frac{\alpha}{2}} * \nu) \hat{\varphi} = C \int |\xi|^{-\frac{\alpha}{2}} \hat{\nu}(\xi) \varphi(\xi) \, d\xi.$$

Since the left hand side is zero for all $\varphi \in S$ we must have $\hat{\nu}(\xi) = 0$ implying $\nu \equiv 0$. For more details, see [29, p. 74]. $\square$

Let us now restrict ourselves to the case of two dimensions. Then we can calculate the Riezs kernels $k_2(x) = 2\pi \log \frac{1}{|x|}$ and $k_1(x) = |x|^{-1}$. We have the following result about the energy integral when $\alpha = 2$.

THEOREM 8.8 (N=2). *Suppose that either*

(i) $\nu(X) = 0$

or

(ii) $supp \, \nu \subset \{x : |x| < 1\}$.

Then $I_2(\nu) = 2\pi \iint \log \frac{1}{|x-y|} \, d\nu(x) \, d\nu(y) \geq 0$ *and* $I_2(\nu) = 0 \Leftrightarrow \nu \equiv 0$.

Proof Also in this theorem we will give two proofs, where the first gives us a slightly weaker result.

PROOF 1.. If $S_\nu \subset \{|x| \leq \frac{1}{2}\}$ then

$$I_2(\nu) = 2\pi \iint \log^+ \frac{1}{|x - y|} \, d\nu(x) \, d\nu(y)$$

and our previous methods will work. $\square$

PROOF 2.. (i) We know that if $\nu(X) = 0$ then $k_2 * \nu = (k_1 * k_1) * \nu$, see [29, p. 51]. Arguing as above gives,

$$I_2(\nu) = \int_{\mathbb{R}^N} (U_1^\nu(z))^2 \, dz.$$

Hence, $I_2(\nu) \geq 0$.

The equality $I_2(\nu) = 0$ implies $U_1^\nu(z) = 0$ a.e. and we continue by arguing as in the proof of Theorem 8.7.

(ii) Let τ be the normalized measure on $\mathbb{T}$, i.e. $d\tau = d\theta/2\pi$, and we define $\nu_1 := \nu - \nu(X)\tau$, where $\nu(X) \neq 0$. Hence, ν_1 is a non-vanishing measure. Thus,

$$U_2^\tau(z) = 2\pi \int_0^{2\pi} \log \frac{1}{|z - e^{i\theta}|} \frac{d\theta}{2\pi} = 0 \text{ for all } |z| < 1.$$

The energy integral is then,

$$I_2(\nu_1) = I_2(\nu) - 2\nu(X)I(\nu,\tau) + \nu(X)^2 I_2(\tau),$$

where $I(\nu,\tau) := \int U_2^\tau(z) \, d\nu(z)$. Since $S_\nu \subset \{|x| < 1\}$ we have that $I(\nu,\tau) = 0$. After a calculation as above we have also, $I_2(\tau) = 0$. Hence $I_2(\nu) = I_2(\nu_1)$ and we conclude $I_2(\nu) \geq 0$. Since ν_1 is non-vanishing, we have always $I_2(\nu) > 0$.

□

Let us now introduce some new notations:

$\mathcal{E}_\alpha$: All signed measures with finite α-energy.

$\mathcal{E}_\alpha^+$: All measures with finite α-energy (no assumptions on compact support).

$\mathfrak{M}_F^+$: All measures with support on the compact set F.

$\overset{\circ}{\mathfrak{M}}{}_F^+$: The set $\{\mu \in \mathfrak{M}_F^+ : \mu(F) = 1\}$.

$\mathcal{E}_F^+$: The set $\mathcal{E}_\alpha^+ \cap \mathfrak{M}_F^+$.

$\overset{\circ}{\mathcal{E}}{}_F^+$: The set $\mathcal{E}_\alpha^+ \cap \overset{\circ}{\mathfrak{M}}{}_F^+$.

We define

$$\|\mu\|^2 := I_\alpha(\mu) = \iint K_\alpha(x - y) \, d\mu(x) \, d\mu(y)$$

for (signed) measures for which the integral converges, i.e. we consider only measures which are such that,

$$\int U_\alpha^{\mu^+} \, d\mu^- < \infty \text{ and } \int U_\alpha^{\mu^-} \, d\mu^+ < \infty.$$

This means that we might have $-\infty < I(\mu) \leq +\infty$. In many cases, we do in fact know that $I(\mu)$ is nonnegative.

If we look at Theorem 8.7, we see that $I_\alpha(\nu) \geq 0$ is true for any signed measure and that $I_\alpha(\nu) = 0$ if and only if $\nu = 0$. It is also clear that $\| \cdot \|$ defines a Hilbert-space structure: we have the triangle inequality and the inner product

(14) $$(\mu, \nu) = \int U_\alpha^\mu \, d\nu.$$

We quote two results [29, p. 90]. Note that there is no assumption on compact supports.

THEOREM 8.9. *The space $\mathcal{E}_\alpha^+$ is complete under the norm (14).*

THEOREM 8.10. *The space $\mathcal{E}_\alpha$ is **not** complete under the norm (14).*

We define $W_\alpha(F) := \inf I_\alpha(\mu)$, $\mu \in \overset{o+}{\mathfrak{M}}_F$. If $\{\mu_n\}$ is a minimizing sequence, i.e.

$$I_\alpha(\mu_n) \to W_\alpha(F),$$

we choose a subsequence, also denoted by $\{\mu_n\}$, such that $\mu_n \to \lambda = \lambda_F$.
By lower semicontinuity,

$$I(\lambda) \le \liminf I(\mu_n) = W_\alpha(F)$$

and we have proved that

$$W_\alpha(F) = I(\lambda_F) \le I(\mu), \ \forall \mu \in \overset{o+}{\mathfrak{M}}_F.$$

We can even prove more: λ_F is in fact the unique minimizer in $\overset{o+}{\mathfrak{M}}_F$. We restrict ourselves to the subclass $\overset{o+}{\mathcal{E}}_F$ of measures with finite α–energy. If $\{\mu_n\}$ is a minimizing sequence in $\overset{o+}{\mathcal{E}}_F$, then $\frac{1}{2}(\mu_n + \mu_m) \in \overset{o+}{\mathcal{E}}_F$ and

$$W_\alpha(F) \le I_\alpha(\frac{1}{2}(\mu_n + \mu_m)) = \|\frac{1}{2}(\mu_n + \mu_m)\|^2.$$

We see that

$$\|\mu_n - \mu_m\|^2 = 2\|\mu_n\|^2 + 2\|\mu_m\|^2 - 4\|\frac{1}{2}(\mu_n + \mu_m)\|^2$$

and thus that $\{\mu_n\}$ is a Cauchy sequence in $\mathcal{E}_F^+$. From Theorem 8.9 and the weak convergence, we see that

$$\|\mu_n - \lambda_F\| \to 0 \text{ as } n \to \infty.$$

The uniqueness of λ_F is proved in a standard way.

DEFINITION 8.11 (THE α–CAPACITY).

$$C_\alpha(F) = 1/W_\alpha(F), \ where$$

$$W_\alpha(F) = \inf\{I_\alpha(\mu) : \mu \in \overset{o+}{\mathfrak{M}}_F\}.$$

Furthermore, if $\overset{o+}{\mathfrak{M}}_F \cap \mathcal{E}_\alpha^+ = \emptyset$, then $W_\alpha(F) = \infty$ and we say that $C_\alpha(F) = 0$.

If $0 < \alpha \le 2$, then this new definition of capacity is the same as the old one for the kernel k_α.

THEOREM 8.12.

$$C_\alpha(F) = 0 \Leftrightarrow \mu(F) = 0,$$

for any measure $\mu \in \mathcal{E}_\alpha^+$.

PROOF. The necessary part.
If $\mu(F) = 0 \ \forall \mu \in \mathcal{E}_\alpha^+$, then $\overset{o+}{\mathcal{E}}_F$ is empty, which implies $C_\alpha(F) = 0$.
The sufficient part.

EXERCISE 8.1. *Show this implication, i.e. show*

$$C_\alpha(F) = 0 \Rightarrow \mu(F) = 0 \ \text{ for all } \mu \in \mathcal{E}_\alpha^+.$$

□

COROLLARY 8.13. *If $C_\alpha(F) = 0$, then $\nu(F) = 0$ for any signed measure $\nu \in \mathcal{E}_\alpha$.*

Note that $I_\alpha(\nu)$ is defined assuming that both $\int U_\alpha^{\nu^+} d\nu^-$ and $\int U_\alpha^{\nu^-} d\nu^+$ are finite.

DEFINITION 8.14 (INNER CAPACITY ZERO). *E is said to have inner capacity zero if for any compact set $F \subset E$, we have $C_\alpha(F) = 0$.*

DEFINITION 8.15 (APPROXIMATELY EVERYWHERE). *Approximately everywhere means except for a set of inner capacity zero.*

8.2. Properties of U_α^λ, where $\lambda = \lambda_F$. In this section, we assume that $C_\alpha(F) > 0$.

(a) $U_\alpha^\lambda(x) \geq W_\alpha(F) = \|\lambda\|^2$ approximately everywhere on F.

PROOF OF (A).. Suppose there exists a compact set $F_0 \subset F$ such that $C_\alpha(F_0) > 0$ and $U_\alpha^\lambda(x) < \|\lambda\|^2$ holds on F_0. Then we can find a measure $\nu \in \overset{\circ}{\mathcal{E}}{}^+_{F_0} = \overset{\circ}{\mathfrak{M}}{}_{F_0} \cap \mathcal{E}_\alpha^+$. The inner product in our Hilbert space will then be,

$$(\nu, \lambda) = \int U_\alpha^\nu \, d\lambda = \int U_\alpha^\lambda \, d\nu < \|\lambda\|^2,$$

keeping in mind that $\nu(X) = 1$.
On the other hand, since for any a, $0 < a \leq 1$,

$$a\nu + (1 - a)\lambda \in \overset{\circ}{\mathcal{E}}{}^+_F,$$

we have

$$\|a\nu + (1 - a)\lambda\| \geq \|\lambda\|.$$

Hence,

$$a^2\|\nu\|^2 + 2a(1 - a)(\nu, \lambda) + (1 - a)^2\|\lambda\|^2 \geq \|\lambda\|^2,$$

giving,

$$2[(\nu, \lambda) - \|\lambda\|^2] + \mathcal{O}(a) \geq 0,$$

i.e.

$$(\nu, \lambda) \geq \|\lambda\|^2.$$

The contradiction proves (a). $\square$

(b) At all points of supp $\lambda \subset F$, we have $U_\alpha^\lambda(x) \leq W_\alpha(F) = \|\lambda\|^2$

PROOF OF (B).. Let $x_0 \in S(\lambda)$ and assume that $U_\alpha^\lambda(x_0) > W_\alpha(F)$. Then due to the lower semicontinuity of $U_\alpha^\lambda(x)$, $U_\alpha^\lambda(x) > W_\alpha(F)$ in a neighbourhood $\mathcal{O}$ of x_0. Moreover $\lambda(\mathcal{O}) > 0$ and there exists a positive number a such that

$$\int_{\mathcal{O}} U_\alpha^\lambda \, d\lambda = (W_\alpha(F) + a)\lambda(\mathcal{O}).$$

Let $\epsilon > 0$ be given. From (a), we know that the closed set

$$T = \{x \in S(\lambda) : U_\alpha^\lambda(x) \leq W_\alpha(F) - \epsilon\}$$

has capacity zero. From Theorem 8.12, we see that $\lambda(T) = 0$. It is now clear that

$$\|\lambda\|^2 = \int_{\mathcal{O}} U_\alpha^\lambda \, d\lambda + \int_{S(\lambda) \backslash \mathcal{O}} U_\alpha^\lambda \, d\lambda \geq$$

$$\geq (W_\alpha(F) + a)\lambda(\mathcal{O}) + (W_\alpha(F) - \epsilon)(1 - \lambda(\mathcal{O})) =$$

$$= W_\alpha(F) + \lambda(\mathcal{O})(a + \epsilon) - \epsilon = \|\lambda\|^2 + \lambda(\mathcal{O})(a + \epsilon) - \epsilon > \|\lambda\|^2,$$

provided that $\epsilon < \frac{\lambda(\mathcal{O})a}{1 - \lambda(\mathcal{O})}$. Hence there can be no such point x_0 and (b) is proved. $\square$

Now, (a) and (b) imply that
$$U_\alpha^\lambda(x) = W_\alpha(F)$$
holds approximately everywhere on the support of λ. Moreover,
$$U_\alpha^\lambda(x) \le W_\alpha(F)$$
everywhere on the support of λ.

8.3. The equilibrium measure. Define $\gamma = \gamma_F := \lambda_F/W_\alpha(F) = C_\alpha(F)\lambda$. γ is called the equilibrium measure on F and $U_\alpha^\gamma(x)$ the equilibrium potential. We see immediately that we have the following properties of the equilibrium measure.

   (a) $U_\alpha^\gamma(x) \ge 1$ approximately everywhere on F.
   (b) $U_\alpha^\gamma(x) = 1$ approximately everywhere on $\mathrm{supp}\,\gamma$.
   (c) $U_\alpha^\gamma(x) \le 1$ everywhere on $\mathrm{supp}\,\gamma$.
   We will also have,
$$\|\gamma\|^2 = I_\alpha(\gamma) = \gamma(F) = C_\alpha(F).$$
For $\alpha \le 2$, $U_\alpha^\gamma(x) = 1$ approximately everywhere on F and $U_\alpha^\gamma(x) \le 1$ everywhere. In this case, we shall say that the measure γ solves the equilibrium problem of Robin. The equilibrium measure may be viewed as the unique solution to a number of variational problems, cf. Theorem 5.7 or [29, p. 138]. We quote two of these problems.

THEOREM 8.16. *Suppose $C_\alpha(F) > 0$. Then the equilibrium measure γ is the unique solution of the following variational problems.*

   (i) *Maximize $\mu(F)$ if $\mu \in \mathfrak{M}_F^+$ and $U_\alpha^\mu(x) \le 1$, $x \in S(\mu)$.*
   (ii) *Among the measures $\mu \in \mathfrak{M}_F^+$ which are such that $\mu(F) = C_\alpha(F)$, find the measure γ for which*
$$\sup_{S(\gamma)} U_\alpha^\gamma(x) = \min \sup_{S(\mu)} U_\alpha^\mu(x).$$

Let us first show that the equilibrium measure γ solves problem (ii). We know that $\sup_{S(\gamma)} U_\alpha^\gamma(x) = 1$. If $\mu(F) = C_\alpha(F)$ and $\sup_{S(\gamma)} U_\alpha^\mu(x) < 1$,
$$\|\mu\|^2 = \int_{S(\gamma)} U_\alpha^\mu(x)\,d\mu(x) < C_\alpha(F) = \|\gamma\|^2,$$
and we deduce that
$$\|\mu/C_\alpha(F)\|^2 < \|\gamma/C_\alpha(F)\|^2 = \|\lambda_F\|^2.$$

But $\mu/C_\alpha(F) \in \overset{o}{\mathfrak{M}}\,{}_F^+$ and $\|\lambda_F\|$ is the minimal value of the norm in this class. This is impossible and we conclude that $\sup_{S(\mu)} U_\alpha^\mu(x) \ge 1$ for all $\mu \in \mathfrak{M}_F^+$ with $\mu(F) = C_\alpha(F)$. Thus the smallest minimum of $\sup_{S(\mu)} U_\alpha^\mu(x)$ is 1. If $\sup_{S(\mu)} U_\alpha^\mu(x) \le 1$, we repeat the previous argument and deduce that
$$\|\mu/C_\alpha(F)\|^2 \le \|\lambda_F\|^2.$$
By, unicity, $\mu/C_\alpha(F) = \lambda_F$ and $\mu = \gamma$.

To prove that the equilibrium measure γ solves problem (i), we consider the functional
$$A(\mu) = \mu(F)/\sup_{S(\mu)} U_\alpha^\mu(x), \quad \mu \in \mathfrak{M}_F^+.$$
If $a > 0$, we have $A(a\mu) = A(\mu)$ and we see that the maximal value of $A(\mu)$ in $\mathfrak{M}_F^+$ is attained for $\mu = a\gamma$ (where $a > 0$) and is equal to $C_\alpha(F)$. Therefore, taking the subclass of $\mathfrak{M}_F^+$ determined by the normalization $\sup_{S(\mu)} U_\alpha^\mu(x) = 1$, we see that the measure γ solves problem (i).

Remark. The following definition of capacity of a compact set is due to *de la Vallée-Poussin:*

$$C_\alpha(F) = \max \mu(F), \quad \mu \in \mathfrak{M}_F^+, \quad U_\alpha^\mu(x) \leq 1, \quad x \in S(\mu).$$

According to problem (i) above, our definition of capacity gives the same result. We note that when $0 < \alpha \leq 2$, the condition $x \in S(\mu)$ is redundant.

Remark. The reader should compare this way of defining capacity with Definition 5.1 and Theorem 8.16 with Theorem 5.7.

8.4. Properties of $C_\alpha(\cdot)$.

DEFINITION 8.17 (INNER CAPACITY). $\underline{C}_\alpha(E) = \sup_{F \subset E} C_\alpha(F)$ *where F is compact.*

(a) $F_1 \subset F_2 \Rightarrow C_\alpha(F_1) \leq C_\alpha(F_2)$.
(b) Denumerable sub-additivity, i.e.

$$\underline{C}_\alpha(\cup F_n) \leq \sum C_\alpha(F_n).$$

Now, let $\epsilon > 0$ and find a compact set $F_0 \subset \cup F_n$ such that $C_\alpha(F_0) \geq \underline{C}_\alpha(\cup F_n) - \epsilon$. Let $\gamma = \gamma_{F_0}$ and $\gamma_n = \gamma|_{F_n}$.

Claim. $\gamma_n(X) \leq C_\alpha(F_n) \sup_{x \in S(\gamma_n)} U_\alpha^{\gamma_n}(x)$.

PROOF OF THE CLAIM.. For any compact F and any $\mu \in \overset{\text{o}+}{\mathcal{E}}_F$ we have,

$$\sup_{x \in S(\mu)} U_\alpha^\mu(x) \geq I_\alpha(\mu) \geq W_\alpha(F).$$

If we take a measure $\nu \in \mathcal{E}_F^+$ and choose μ as $\nu/\nu(F)$ which is in $\overset{\text{o}+}{\mathcal{E}}_F$ then we will have,

$$\nu(F) \leq C_\alpha(F) \sup_{x \in S(\nu)} U_\alpha^\nu(x).$$

Choosing $\nu = \gamma_n$, we obtain the claim. $\quad\square$

PROOF OF (B)..

$$\underline{C}_\alpha(\cup F_n) - \epsilon \leq C_\alpha(F_0) = \gamma(F_0) \leq \sum \gamma_n(X) \leq$$

$$\leq \sum C_\alpha(F_n) \sup_{x \in S(\gamma_n)} U_\alpha^{\gamma_n}(x) \leq$$

$$\leq \sum C_\alpha(F_n) \sup_{x \in S(\gamma)} U_\alpha^\gamma(x) \leq \sum C_\alpha(F_n)$$

$\square$

(c) $C_\alpha(\cdot)$ is *continuous on the right*, i.e. for any $\epsilon > 0$ there exists a neighbourhood $\mathcal{O}$ of F such that for all compact sets F', $F \subset F' \subset \mathcal{O}$, we have $C_\alpha(F') \leq \epsilon + C_\alpha(F)$.

PROOF. Assume the contrary, i.e. that there is an $\epsilon > 0$ such that for $F_1 \supset F_2 \supset \ldots \supset F$, $\cap_1^\infty F_n = F$, we have

$$C_\alpha(F_n) \geq \epsilon + C_\alpha(F).$$

Let $\gamma_n = \gamma_{F_n}$ be the equilibrium measure for F_n. It is clear that

$$I_\alpha(\gamma_n) = \gamma_n(X) = C_\alpha(F_n).$$

Let now $\{\gamma_n\}$ be a weakly convergent subsequence $\gamma_n \to \mu \in \mathfrak{M}_F^+$. (We do not change the notation when going to the subsequence.) We end up with

$$I_\alpha(\mu) \leq \liminf_{n \to \infty} I_\alpha(\gamma_n).$$

The total mass is then

$$\mu(F) = \lim \gamma_n(X) \geq C_\alpha(F) + \epsilon$$

Since $\mu/(C_\alpha(F) + \epsilon)$ has total mass ≥ 1 and $1/C_\alpha(F) = \inf I_\alpha(\gamma)$ for $\gamma \in \overset{\circ}{\mathfrak{M}}_F^+$, we see that

$$I_\alpha(\mu) = \int U_\alpha^\mu(x)\, d\mu(x) \geq (C_\alpha(F) + \epsilon)^2/C_\alpha(F) \geq C_\alpha(F) + 2\epsilon.$$

Since $C_\alpha(F_n)$ is decreasing,

$$C_\alpha(F_n) \geq I_\alpha(\mu) \geq C_\alpha(F) + 2\epsilon.$$

Repeating this argument p times will give us

$$C_\alpha(F_n) \geq C_\alpha(F) + 2^p\epsilon \to \infty, \text{ as } p \to \infty$$

which is the desired contradiction. $\square$

(d) If $0 < \alpha \leq 2$ then $C_\alpha(\cdot)$ is *strongly subadditive*[6], i.e.

$$C_\alpha(F_1 \cup F_2) + C_\alpha(F_1 \cap F_2) \leq C_\alpha(F_1) + C_\alpha(F_2).$$

Landkof calls this property "convexity". To prove that (d) holds, we need the following theorem.

THEOREM 8.18 (THE SECOND MAXIMUM PRINCIPLE). *Suppose $\mu \in \mathcal{E}_\alpha^+$ and that λ is any measure. Then if $U_2^\mu(x) \leq U_2^\lambda(x)$ μ-a.e., it holds everywhere in $\mathbb{R}^N$.*

Remark. This can be generalized: for $0 < \alpha \leq 2$,

$$U_\alpha^\mu(x) \leq U_\alpha^\lambda(x) \ \mu\text{-a.e.} \ \Rightarrow U_\alpha^\mu(x) \leq U_\alpha^\lambda(x)$$

everywhere in $\mathbb{R}^N$ (see for example [29, p. 115]). The proof of Theorem 8.18 is long and will be given in the next section.

EXERCISE 8.2. *Give an estimate of $C_\alpha(B_R)$ for different values of α. Here, B_R is a ball of radius R. Discuss the support of the equilibrium measure.*

EXERCISE 8.3. *Find $C_2(B_R)$ and the corresponding equilibrium measure.*

EXERCISE 8.4. *Estimate $C_\alpha(F)$ when F is F_1 or F_2 where*

$$F_1 = \{x \in \mathbb{R}^N : |x| \leq 1, x_N = 0\},$$

$$F_2 = \{x \in \mathbb{R}^N : |x| \leq 1, x_N = x_{N-1} = 0\}.$$

EXERCISE 8.5. *Let $F = \{x \in \mathbb{R}^2 : x_2 = 0, x_1 \in E_C\}$, where E_C is the $\frac{1}{3}$-Cantor set on the interval $[0, 1]$. Estimate $C_\alpha(F)$.*

EXERCISE 8.6. *For what values of α is it true that $|F| = 0$ if $C_\alpha(F) = 0$? Here $|\cdot|$ denotes Lebesgue measure in $\mathbb{R}^N$.*

---

[6]The terminology is from Doob [10, p. 752].

EXERCISE 8.7. *Let $M_\alpha(\cdot)$ denote Hausdorff measure defined with respect to the function $h(r) = r^\alpha$. Let F be a subset of $\mathbb{R}^{N-1}$ and assume that we know something about $M_\alpha(F)$. Imbedding R^{N-1} in $\mathbb{R}^N$ in a natural way, discuss what we can say about $C_\beta(F)$ where the capacity is defined for subsets of $\mathbb{R}^N$.*

Solve the problems for those values of the dimension N for which they make sense.

8.5. Potentials of measures in the whole space. We will in this section consider only $N \geq 3$. We will not assume that supports are compact.

Notation. Let us write $\fint_{|x|=R} f(x)\, d\sigma_x$ for the mean value

$$\frac{1}{\int_{|x|=R} d\sigma_x} \int_{|x|=R} f(x)\, d\sigma_x.$$

LEMMA 8.19.

(15)
$$\fint_{|x|=R} \frac{d\sigma_x}{|x-y|^{N-2}} = \begin{cases} |y|^{2-N} & \text{if } |y| \geq R, \\ R^{2-N} & \text{if } |y| < R. \end{cases}$$

PROOF. EXERCISE 8.8. *Prove this!*
□

LEMMA 8.20. *Let μ be a measure in $\mathbb{R}^N$. A necessary and sufficient condition for finiteness a.e. of the potential $U_\alpha^\mu(x)$ is the inequality*

(16)
$$\int_{|y|>1} |y|^{\alpha-N} d\mu(y) < \infty.$$

PROOF OF LEMMA 8.20.. Assume that (16) holds. Then for $R > 1$

$$\int_{|x|<R} U_\alpha^\mu(x)\, dx = \int d\mu(y) \int_{|x|<R} |x-y|^{\alpha-N} dx =$$

$$= \left(\int_{|y|<2R} + \int_{|y|>2R} \right) d\mu(y) \int_{|x|<R} |x-y|^{\alpha-N} dx = I_1 + I_2.$$

Let us estimate the two terms separately,

$$I_1 \leq \int_{|y|<2R} d\mu(y) \int_{|z|<3R} |z|^{\alpha-N} dz =$$

$$= C_N \int_{|y|<2R} d\mu(y) \int_0^{3R} r^{\alpha-1} dr =$$

$$= \frac{C_N}{\alpha} (3R)^\alpha \int_{|y|<2R} d\mu(y).$$

Since $|x - y| \geq |y| - |x| \geq |y|/2$ if $|y| > 2R$ and $|x| < R$, we have

$$I_2 \leq \int_{|y|>2R} d\mu(y) \int_{|x|<R} |y|^{\alpha-N} dx \cdot 2^{N-\alpha} =$$

$$= C_N 2^{N-\alpha} RN \int_{|y|>2R} |y|^{\alpha-N} d\mu(y) < \infty.$$

We conclude that $U_\alpha^\mu \in L^1_{loc}(\mathbb{R}^N)$ and thus U_α^μ must be finite a.e.

Conversely, assume that (16) does not hold . Then

$$\int_{\mathbb{R}^N} |x-y|^{\alpha-N} d\mu(y) \geq \int_{|y|>2|x|} |x-y|^{\alpha-N} d\mu(y) \geq$$

$$\geq \int_{|y|>2|x|} (2|y|)^{\alpha-N} d\mu(y) = \infty$$

and the potential is infinite. □

In view of the above lemma, let us assume that $\int_{|y|>1} |y|^{2-N} d\nu(y) < \infty$ and let $U_2^\nu(x) = \int C_N |x - y|^{2-N} d\nu(y)$.

Let us estimate

$$\fint_{|x|=R} U_2^\nu(x) \, d\sigma_x = \int d\nu(y) \fint_{|x|=R} \frac{C_N d\sigma_x}{|x - y|^{N-2}} =$$

$$\int_{|y|<R} C_N \, d\nu(y) R^{2-N} + \int_{|y|\geq R} C_N \, d\nu(y) |y|^{2-N}.$$

We will then have

$$\int_{|y|<R} C_N \, d\nu(y) R^{2-N} = R^{2-N} C_N \int_{|y|<A} d\nu(y) +$$

$$+ C_N \int_{A\leq|y|<R} d\nu(y) R^{2-N} \leq$$

$$\leq R^{2-N} C_N \int_{|y|<A} d\nu(y) + C_N \int_{|y|\geq A} |y|^{2-N} d\nu(y) \to 0 + 0,$$

as $R \to \infty$ and thereafter $A \to \infty$,

and trivially

$$\int_{|y|\geq R} C_N \, d\nu(y) |y|^{2-N} \to 0 \quad \text{as } R \to \infty.$$

We have proved that

$$\fint_{|x|=R} U_2^\nu(x) \, d\sigma_x \to 0 \quad \text{as } R \to \infty.$$

THEOREM 8.21 (LANDKOF, P. 106). *A superharmonic function f in $\mathbb{R}^N, N \geq 3$, is a Newtonian potential of a measure in $\mathbb{R}^N$ if and only if f is non-negative and $\fint_{|x|=R} f(x) \, d\sigma_x \to 0$ as $R \to \infty$.*

PROOF. The first part is done above for $f = U_2^\nu$.

For the second part of the proof let us assume that the mean values tend to zero as $R \to \infty$ and let the measure μ be the *Riesz mass* of f, i.e. $\mu = -\frac{1}{4\pi^2}\Delta f$. Let μ_r be the restricted measure $\mu_r = \mu|_{\{|x|\leq r\}}$. Furthermore, let $F_r(x) := f(x) - U_2^{\mu_r}(x)$. It follows from the construction that F_r is harmonic in $\{|x| \leq r\}$, see [29, p. 101].

We now claim that F_r is superharmonic in $\mathbb{R}^N$. In fact, if $|x_1| \geq r$, we write

$$f(x) = U_2^{\mu_{r_1}}(x) + h_1(x)$$

where $r_1 > |x_1|$ and $h_1(x)$ is harmonic in $\{|x| \leq r_1\}$. If $\nu = \mu_{r_1} - \mu_r$, we see that

$$F_r(x) = f(x) - U_2^{\mu_r}(x) = U_2^\nu(x) + h_1(x).$$

The right hand side is superharmonic near x_1, see Theorem 8.3, which proves our claim. Hence

$$\fint_{|x|=R} F_r(x) \, d\sigma_x \to 0 \quad \text{as } R \to \infty.$$

It follows that

$$f(x_0) - U_2^{\mu_r}(x_0) \geq \fint_{|x-x_0|=R} F_r(x) \, d\sigma_x \to 0 \quad \text{as } R \to \infty.$$

Hence $F_r(x_0) \geq 0, \forall x_0 \in \mathbb{R}^N$ Letting $r \to \infty$ will then give us $F(x) := f(x) - U_2^\mu(x) \geq 0$, and harmonic in $\mathbb{R}^N$. $F(x)$ must therefore be constant and the condition above that

$$\fint_{|x|=R} F_r(x)\, d\sigma_x \to 0 \quad \text{as } R \to \infty$$

gives us that this constant must be zero. That is, $f(x) = U_2^\mu(x)$ and the proof is finished. $\square$

COROLLARY 8.22. *If $f \geq 0$ and f is superharmonic in $\mathbb{R}^N$ and majorized by a potential of a measure ν, $f(x) \leq U_2^\nu(x)$, then is $f = U_2^\mu$ for some measure $\mu \geq 0$.*

COROLLARY 8.23. *If f is a non-negative superharmonic function in $\mathbb{R}^N$, then $f(x) = U_2^\mu(x) + \alpha$, where $\alpha = \lim_{R\to\infty} \fint_{|x|=R} f(x)\, d\sigma_x$.*

PROOF. Let $f_1(x) = f(x) - \alpha \geq 0$. We know that f_1 is superharmonic and that $\fint_{|x|=R} f_1(x)\, d\sigma_x \to 0$. Applying Theorem 8.21, we conclude that $f_1 = U_2^\mu$. $\square$

THEOREM 8.24. *Let $x \in \mathbb{R}^N$. A function f admits a representation $f = U_2^\mu + h$, where μ is a measure and h is a harmonic function, if and only if f is superharmonic in $\mathbb{R}^N$ and has a harmonic minorant.*

PROOF. We first prove the necessary part: if f is superharmonic in $\mathbb{R}^N$ and h_1 is a harmonic minorant, then $f - h_1$ is nonnegative and superharmonic. Corollary 8.23 implies $f - h_1 = U_2^\mu + \alpha$. Put $h = \alpha + h_1$. We immediately get $h \geq h_1$, in fact, h is the greatest harmonic minorant of f.

Conversely, if we have $f = U_2^\mu + h$, we have $f \geq h$. Since we have assumed that h is harmonic then h is also a harmonic minorant to f. $\square$

THEOREM 8.25. *A superharmonic function f in $\mathbb{R}^N$ has a harmonic minorant if and only if*

$$\lim_{R\to\infty} \fint_{|x|=R} f\, d\sigma_x > -\infty.$$

PROOF. We show first that the condition is necessary. If f has a harmonic minorant, then f is of the form $U_2^\mu + h$ (cf. Theorem 8.24) and the limit of the mean values is $h(0) > -\infty$.

Conversely, we have from the Poisson integral that

$$H_R(x) = \frac{C_N}{R} \int_{|y|=R} f(y) \frac{R^2 - |x|^2}{|x-y|^N}\, d\sigma_y.$$

We know that $H_R = f$ on ∂B_R and that H_R is the greatest harmonic minorant of f in $\{|x| < R\}$.

Let us now take $R_1 < R$. H_R is a harmonic minorant of f in $\{|x| < R_1\}$ and H_{R_1} is the greatest harmonic minorant of f in $\{|x| < R_1\}$. Thus, $H_R \leq H_{R_1}$. Hence $\lim_{R\to\infty} H_R$ exists and

$$\lim_{R\to\infty} H_R = \begin{cases} -\infty & \quad \text{or} \\ \text{harmonic in } \mathbb{R}^N. \end{cases}$$

By assumption,

$$H_R(0) = \fint_{|y|=R} f(y)\, d\sigma_y \nrightarrow -\infty,$$

hence $h = \lim_{R\to\infty} H_R$ is harmonic in $\mathbb{R}^N$ and is a harmonic minorant of f. $\square$

THEOREM 8.26. *Suppose we have an increasing sequence of potentials $\{U_2^{\mu_n}\}$ where the μ_n:s are measures and suppose that there exists a measure ν such that $U_2^{\mu_n} \leq U_2^{\nu}$, $\forall n$, then there is a measure μ such that*

$$\lim_{n \to \infty} U_2^{\mu_n}(x) = U_2^{\mu}(x)$$

and $\mu_n \rightharpoonup \mu$.

PROOF. Let $g(x) := \lim_{n \to \infty} U_2^{\mu_n}(x)$.

Claim. g is superharmonic.

Since $U_2^{\mu_n}$ is superharmonic

$$U_2^{\mu_n}(x_0) \geq \oint_{|x|=r} U_2^{\mu_n}(x_0 + x)\, d\sigma_x.$$

This is true for all $r \geq 0$. Letting n tend to infinity we will, by monotone convergence, obtain

$$g(x_0) \geq \oint_{|x|=r} g(x_0 + x)\, d\sigma_x.$$

Hence, g satisfies the superharmonic mean value property and it is easy to see that g is also semi continuous from below. The claim is proved.

We note that $0 \leq g \leq U_2^{\nu}$ and that g is superharmonic. Corollary 8.22 gives then the existence of a measure μ such that $g = U_2^{\mu}$.

For the proof of the last part, i.e. $\mu_n \rightharpoonup \mu$, we recall that for any signed measure λ we have

$$\int U_2^{\mu_n}\, d\lambda \to \int U_2^{\mu}\, d\lambda, \text{ or}$$

$$\int U_2^{\lambda}\, d\mu_n \to \int U_2^{\lambda}\, d\mu.$$

If $\varphi \in C_0^{\infty}(\mathbb{R}^N)$ then $\varphi(x) = -k_2 * \Delta\varphi \frac{1}{4\pi^2} = U_2^{\lambda}$ if $\lambda = -\frac{\Delta\varphi}{4\pi^2}$, see Lemma 1 in [29, p. 74]. Hence

$$\int \varphi\, d\mu_n \to \int \varphi\, d\mu, \quad \forall \varphi \in C_0^{\infty}(\mathbb{R}^N).$$

The weak convergence is proved. $\square$

Let us now state a rather interesting result.

THEOREM 8.27. *Suppose that μ is a measure with $I_2(\mu) < \infty$ and that f is superharmonic and non–negative in $\mathbb{R}^N$. If*

(17) $$U_2^{\mu}(x) \leq f(x)$$

holds μ–a.e., then the inequality holds everywhere.

PROOF. Set $V(x) = \min\{U_2^{\mu}(x), f(x)\}$ to get $V(x)$ superharmonic and $V(x) \leq U_2^{\mu}(x)$. Then we know from 8.22 that there is a measure λ such that $V = U_2^{\lambda}$.

If $\lambda = \mu$, then

$$U_2^{\lambda}(x) = U_2^{\mu}(x) = \min\{U_2^{\mu}(x), f(x)\},$$

i.e. $U_2^{\mu}(x) \leq f(x)$ everywhere. Let us now show that this must be the case, i.e. $\lambda = \mu$

The energy integral of λ is

$$I_2(\lambda) = \int U_2^{\lambda}\, d\lambda \leq \int U_2^{\mu}\, d\lambda =$$

$$= \int U_2^{\lambda}\, d\mu \leq \int U_2^{\mu}\, d\mu = I_2(\mu) < \infty.$$

Hence $I(\lambda) < \infty$. Let

$$E := \{x : U_2^\mu(x) > f(x)\} = \{x : U_2^\mu(x) > U_2^\lambda(x)\}.$$

On $\complement E$, we have $U_2^\mu(x) = U_2^\lambda(x)$. It follows that

$$\|\mu - \lambda\|^2 = \int_{\mathbb{R}^N} (U_2^\mu - U_2^\lambda)(d\mu - d\lambda) = \int_E (U_2^\mu - U_2^\lambda)(d\mu - d\lambda).$$

We have assumed that the inequality (17) holds μ–a.e. and that implies

$$\|\mu - \lambda\|^2 = -\int_E (U_2^\mu - U_2^\lambda)\, d\lambda.$$

Since $U_2^\mu \geq U_2^\lambda$ we get

$$0 \leq \|\mu - \lambda\|^2 \leq 0$$

Lemma 5.10 gives us finally $\mu = \lambda$ □

We can now easily prove the second maximal principle, Theorem 8.18, by letting $f(x)$ be $U_2^\sigma(x)$, where σ is a measure. It follows immediately from Theorem 8.27 that

$$U_2^\mu \leq U_2^\sigma \quad \mu\text{– a.e.} \;\Rightarrow U_2^\mu \leq U_2^\sigma \quad \text{everywhere.}$$

EXERCISE 8.9. *Let $\Phi(x) = \Phi(|x|)$ be a fundamental solution of Laplace's equation in $\mathbb{R}^2$. Compute*

$$\fint_{|x|=R} \Phi(x - y)\, d\sigma_x, \;\; y \in \mathbb{R}^2.$$

8.6. The Green potential. Let Ω be an open set in $\mathbb{R}^N$. Assume that there exists a Green function in Ω. The Green function is constructed in the following way. Let $\Phi(x - x_0)$ be the fundamental solution of Laplace's equation, where $r = |x|$,

$$\Phi(x) = \phi(r) = \begin{cases} \log \frac{1}{r} & \text{if } N = 2 \\ r^{2-N} & \text{if } N \geq 3, \end{cases}$$

and solve the Dirichlet problem in Ω

$$\begin{cases} \Delta u & = & 0, & x \in \Omega^\circ \\ u|_{\partial\Omega} & = & \Phi(x - x_0) & x \in \partial\Omega. \end{cases}$$

The Green function is defined as

$$G(x, x_0) := \Phi(x - x_0) - u(x)$$

and we can define the corresponding potential.

DEFINITION 8.28 (GREEN POTENTIAL).

$$G\mu(x) = \int G(x, y)\, d\mu(y),$$

where the support of μ is in $\bar{\Omega}$.

We will now state, without proof, a celebrated theorem of F. Riesz.

THEOREM 8.29 (THE RIESZ REPRESENTATION THEOREM). *Let Ω be as above and let u be superharmonic in Ω such that either $u \geq 0$ or u has a harmonic minorant in Ω. Then*

$$u = G\mu + h,$$

where $G\mu$ is the Green potential of a measure μ and h is the greatest harmonic minorant of u in Ω.

See [26, p. 116] for the proof.

COROLLARY 8.30. *A non–negative superharmonic function in* Ω *is a Green potential if and only if its greatest harmonic minorant is zero.*

Example. Let u be superharmonic, non–negative in the unit disk in the plane and $u = 0$ on $\mathbb{T}$, then $u = G\mu$.

μ is called the *Riesz mass* of the superharmonic function u.

EXERCISE 8.10. *The Poisson–Jensen formula in the unit disc in the plane is an example of the Riesz decomposition theorem. Explain the connection.*

8.7. Strong Subadditivity. We repeat the definition of this property.

DEFINITION 8.31 (STRONG SUBADDITIVITY). $C_\alpha(\cdot)$ *is strongly subadditive if*

$$C_\alpha(F_1 \cap F_2) + C_\alpha(F_1 \cup F_2) \le C_\alpha(F_1) + C_\alpha(F_2).$$

We will in this section show that $C_2(\cdot)$ is strongly subadditive. This can be generalized to C_α where $0 < \alpha \le 2$.

Recall that we defined *approximately everywhere* (app.e.) to be *except for a set of inner capacity zero*. Let γ be the equilibrium distribution of a compact set. We then showed that

(1) $U_2^\gamma(x) = 1$ approximately everywhere on the support $S(\gamma)$.
(2) $U_2^\gamma(x) \le 1$ everywhere.

Claim.

$$\mu(\{x \in S(\gamma) : U_\alpha^\gamma(x) < 1\}) = 0 \quad \forall \mu \in \mathcal{E}_\alpha^+.$$

PROOF OF THE CLAIM.. Let $\epsilon > 0$ be given and define the compact set

$$E_\epsilon := \{x \in S(\gamma) : U_\alpha^\gamma(x) \le 1 - \epsilon\}.$$

E_ϵ is a subset of a set of inner capacity zero and by using Theorem 8.12 we conclude that $\mu(E_\epsilon) = 0$ and

$$\mu(E) = \lim_{\epsilon \searrow 0} \mu(E_\epsilon) = 0$$

□

LEMMA 8.32. *Let* ν *be a signed measure with compact support and with* $I_\alpha(\nu) < \infty$. *If* $U_\alpha^\nu(x) \ge 0$ *approximately everywhere on the support* $S(\nu)$, *then* $\nu(\mathbb{R}^N) \ge 0$.

PROOF. Let γ be the equilibrium measure on $S(\nu)$, then

$$0 \le \int U_\alpha^\nu \, d\gamma = \int U_\alpha^\gamma \, d\nu = \nu(\mathbb{R}^N)$$

since $U_\alpha^\gamma = 1$ approximately everywhere on $S(\nu)$.   □

THEOREM 8.33. $C_2(\cdot)$ *is strongly subadditive.*

PROOF. Let $h(F)$ be the equilibrium potential for the compact set F. If $F' \subset F$ then we have that if

(1) $h(F) = h(F')$ approximately everywhere on $S(\gamma_{F'})$, i.e. $\gamma_{F'}$–a.e.
    then the second maximum principle implies
(2) $h(F) \ge h(F')$ everywhere.

We repeatedly use the two properties above on the expression:

$$h(F_1) + h(F_2) - h(F_1 \cup F_2) - h(F_1 \cap F_2) =$$

$$= \begin{cases} h(F_1) - h(F_1 \cap F_2) - (h(F_1 \cup F_2) - h(F_2)) & \geq 0 \text{ app.e. on } F_2, \\ h(F_2) - h(F_1 \cap F_2) - (h(F_1 \cup F_2) - h(F_1)) & \geq 0 \text{ app.e. on } F_1. \end{cases}$$

Thus

$$h(F_1) + h(F_2) \geq h(F_1 \cup F_2) + h(F_1 \cap F_2) \quad \text{app.e. on } S(\gamma_{F_1 \cup F_2}).$$

The second maximum principle gives then

$$h(F_1) + h(F_2) \geq h(F_1 \cup F_2) + h(F_1 \cap F_2) \quad \text{everywhere.}$$

Finally, letting ν, defined by $d\nu := d\gamma_{F_1} + d\gamma_{F_2} - d\gamma_{F_1 \cup F_2} - d\gamma_{F_1 \cap F_2}$, be the signed measure in Lemma 8.32, then we conclude that

$$C_2(F_1) + C_2(F_2) \geq C_2(F_1 \cup F_2) + C_2(F_1 \cap F_2)$$

$\square$

Remark. If we use Landkof's Theorem 1.31 to generalize the proof of Theorem 8.27 to get the generalized second maximum principle , we can generalize Theorem 8.33 above to hold for C_α, where $0 < \alpha \leq 2$.

We define inner and outer capacity:

$$\underline{C}_\alpha(E) := \sup_{F \subset E} C_\alpha(F),$$

$$\bar{C}_\alpha(E) := \inf_{\mathcal{O} \supset E} \underline{C}_\alpha(\mathcal{O}),$$

where F is compact and $\mathcal{O}$ is open.

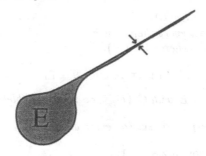

FIGURE 8.1. If the "spine" is narrow enough, then the capacity can be finite.

It is possible to construct a set E with infinite diameter in $\mathbb{R}^N$, $N \geq 3$, and with finite inner capacity if we make E "thin" enough far away from the origin, see picture 8.1.

DEFINITION 8.34 (CAPACITABILITY). *A set E is capacitable if $\underline{C}_\alpha(E) = \bar{C}_\alpha(E)$.*

If $\mathcal{O}$ is an open set, then

$$\underline{C}_\alpha(\mathcal{O}) = \sup_{F \subset \mathcal{O}} C_\alpha(F) \quad \text{and}$$

$$\bar{C}_\alpha(\mathcal{O}) = \inf_{\mathcal{O}' \supset \mathcal{O}} \underline{C}_\alpha(\mathcal{O}') = \underline{C}_\alpha(\mathcal{O}).$$

Hence, open sets are capacitable.

What about compact sets?

$$\underline{C}_\alpha(F) = C_\alpha(F) \quad \text{and}$$

$$\bar{C}_\alpha(F) = \inf_{\mathcal{O} \supset F} \underline{C}_\alpha(\mathcal{O}) = C_\alpha(F).$$

The last equality is true because $C_\alpha(\cdot)$ is continuous on the right. We have proved the following theorem.

THEOREM 8.35. *Compact and open sets are capacitable.*

EXERCISE 8.11. *Prove that we have strong subadditivity also for open sets, i.e.*

$$C_\alpha(\mathcal{O}_1 \cap \mathcal{O}_2) + C_\alpha(\mathcal{O}_1 \cup \mathcal{O}_2) \leq C_\alpha(\mathcal{O}_1) + C_\alpha(\mathcal{O}_2).$$

We assume that the corresponding relation holds when the open sets are replaced by compact sets.

THEOREM 8.36. *Assume* $\underline{C}_\alpha(E) < \infty$. *Then there is a unique measure* $\gamma = \gamma_E$, *with support in* $\bar{E}$ *such that*

(a) $\|\gamma\|^2 = \gamma(\mathbb{R}^N) = \underline{C}_\alpha(E)$.
(b) $U_\alpha^\gamma(x) \geq 1$ *approximately everywhere on* E.
(c) $U_\alpha^\gamma(x) \leq 1$ *everywhere on* $S(\gamma)$.

See Theorem 2.6 on page 145 in [29] for the proof.

THEOREM 8.37. *Assume* $\bar{C}_\alpha(E) < \infty$. *Then there is a unique measure* $\gamma^* = \gamma_E^*$, *with support in* $\bar{E}$ *such that*

(a) $\|\gamma^*\|^2 = \gamma^*(\mathbb{R}^N) = \bar{C}_\alpha(E)$.
(b) $U_\alpha^{\gamma^*}(x) \geq 1$ *quasi everywhere[7] on* E.
(c) $U_\alpha^{\gamma^*}(x) \leq 1$ *everywhere on* $S(\gamma^*)$.

PROOF. See Theorem 2.7 on page 149 in [29]. □

THEOREM 8.38. *A set* E *with* $\bar{C}_\alpha(E) < \infty$ *is capacitable if and only if* $\gamma_E = \gamma_E^*$.

THEOREM 8.39. *Every Borel set (or every analytic set) in* $\mathbb{R}^N$ *is capacitable.*

PROOF. See [29, Theorem 2.8]. □

Let us now state two key facts about summability of inner and outer capacity. For the verification, see [29].

• If $\{E_n\}$ is a sequence of Borel sets, then

$$\underline{C}_\alpha(\cup_1^\infty E_n) \leq \sum_1^\infty \underline{C}_\alpha(E_n).$$

• Let $\{E_n\}$ be a sequence of *arbitrary* sets, then

$$\bar{C}_\alpha(\cup_1^\infty E_n) \leq \sum_1^\infty \bar{C}_\alpha(E_n).$$

---

[7]quasi everywhere = except for a set of outer capacity zero.

LEMMA 8.40. *If $E_1 \subset E_2 \subset E_3 \subset \ldots$, and $E = \cup_1^\infty E_n$, then*

$$\bar{C}_\alpha(E) = \lim_{n \to \infty} \bar{C}_\alpha(E_n).$$

If the sets in the sequence $\{E_n\}$ are Borel sets, then

$$\underline{C}_\alpha(E) = \lim_{n \to \infty} \underline{C}_\alpha(E_n).$$

For the proof of the lemma see [29, p. 154].

COROLLARY 8.41 (OF THEOREM 8.39). *If E is a Borel set with positive capacity, $C_\alpha(E) > 0$, then there exists a compact subset, $F \subset E$, with $C_\alpha(F) > 0$.*

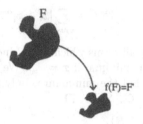

FIGURE 8.2. The capacity decreases under contraction.

8.8. Metric properties of capacity. A function f is defined to be a *contraction* if for all $x, y \in \mathbb{R}^N$, $|f(x) - f(y)| \leq |x - y|$, see figure 8.2.

THEOREM 8.42. *If $f : F \to F' \in \mathbb{R}^N$ is a contraction, then $F' = f(F)$ is also compact and $C_\alpha(F') \leq C_\alpha(F)$.*

8.9. The support of the equilibrium measures. Let us now study the location of the support of the equilibrium measure. We will prove that if $\alpha \geq 2$, then the measure has support essentially on the "outer" boundary. Consider a compact set F, not necessary simply connected, with the complement, $\complement F = G_\infty \cup \bigcup_{i \geq 1} G_i$, where the G_i:s are the components of the complement and G_∞ is the unbounded component, i.e. $\infty \in G_\infty$. Let

$$\bar{\bar{F}} := \complement G_\infty = F \cup \bigcup_{i \geq 1} G_i$$

and $S := \partial \bar{\bar{F}}$, see figure 8.3.

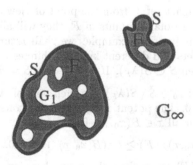

FIGURE 8.3. The support of the equilibrium measure lies on the "thick" S boundary.

THEOREM 8.43. *If $\alpha \geq 2$, then*

$$C_\alpha(F) = C_\alpha(\bar{\bar{F}}) = C_\alpha(S).$$

PROOF. We will prove the second equality using the notion of generalized diameter, see definition 7.2. Then $\inf_{x_i \in F} D_n(F)$ is assumed at $\{x_\nu\}_1^n$. Let us now put the mass $1/n$ at each point and define the measure μ_n.

We know that the unique measure μ that minimizes the energy integral I_α for measures in $\overset{o+}{\mathfrak{M}_F}$ is the weak limit of a subsequence of $\{\mu_n\}$.

If supp $\mu_n \subset S$ $\forall n$, then supp $\mu \subset S$.

Consider the sum

$$\sum_{1 \leq i < j \leq n}^{(\nu_0)} |x_i - x_j|^{\alpha - N} + \sum_{i \neq \nu_0} |x_i - x|^{\alpha - N}.$$

Here $\sum^{(\nu_0)}$ means summation over all terms which do not include the index ν_0. By the definition of D_n the second sum is minimal if $x = x_{\nu_0}$. We also note that the second sum is superharmonic telling us that the minimum is only attained on the boundary S. Thus we have proved $C_\alpha(\bar{\bar{F}}) = C_\alpha(S)$. $\square$

Example. Let $F = \bar{B}_R = \{|x| \leq R\}$.

- For $\alpha \geq 2$, the support of λ_F is a subset of $\{|x| = R\}$. By symmetry, λ_F is uniformly distributed over $\{|x| = R\}$ and it is possible to compute a constant only depending on N and α such that

$$C_\alpha(\bar{B}_R) = R^{N-\alpha}/C(N, \alpha).$$

- For $\alpha < 2$ we have a different picture. The support of $\lambda_F = \{|x| \leq R\}$ and λ_F is absolutely continuous with density λ_F which can in fact be explicitly calculated:

$$\lambda_F(x) = A(R^2 - |x|^2)^{-\alpha/2},$$

hence

$$C_\alpha(\bar{B}_R) = R^{N-\alpha}/C_1(N, \alpha).$$

Let us now turn to the case $0 < \alpha \leq 2$. Let F be a given compact set and define $\check{F}$.

DEFINITION 8.44 (REDUCED SET OF F).

$$\check{F} = \{x \in F : C_\alpha(\mathcal{O} \cap F) > 0 \text{ for each neighbourhood } \mathcal{O} \text{ of } x.\}.$$

The essential part of a compact set F from the point of view of potential theory is the reduced set $\check{F}$. If there are isolated points in F, they will all be in the remaining exceptional set $F \setminus \check{F}$. This is just one example: we shall return to these questions later when discussing polar sets and different kinds of thinness.

Claim. Let $0 < \alpha \leq 2$, then $\check{S} \subset S(\lambda_F)$. If $\alpha = 2$, then $\check{S} = S(\lambda_F)$.

PROOF OF THE CLAIM.. If $x_0 \in \check{S} \setminus S(\lambda_F)$ it is then possible to find a ball $B(x_0, r)$ such that $B(x_0, r) \cap S(\lambda_F) = \emptyset$. The potential $U_\alpha^{\lambda_F}(x)$ is not constant in $G_\infty \cup B(x_0, r)$ implying $U_\alpha^{\lambda_F}(x) < W_\alpha(F)$ for all $x \in B(x_0, r)$. But

$$C_\alpha(B(x_0, r) \cap F) \geq C_\alpha(B(x_0, r) \cap S) > 0.$$

We also have $U_\alpha^\lambda(x) = W_\alpha(F)$ approximately everywhere on F, see [29, p. 137]. Therefore, $B(x_0, r)$ contains a point x where $U_\alpha^{\lambda_F}(x) = W_\alpha(F)$. This is wrong and we conclude $\check{S} \subset S(\lambda_F)$.

If $\alpha = 2$ and if $x_0 \in S(\lambda_F)$, then for all small $r > 0$, we have $\lambda_F(B(x_0, r) \cap S) = \lambda_F(B(x_0, r)) > 0$. This implies $C_\alpha(B(x_0, r) \cap S) > 0$. Hence $x_0 \in \check{S}$ and we have $\check{S} = S(\lambda_F)$ in this case. $\square$

Example.

(1) Let $0 < \alpha < N$ and let $f(x) := x/a$ where $a > 1$. Then $C_\alpha(f(F)) = a^{\alpha-N}C_\alpha(F)$.

(2) Let f be a projection, then $C_\alpha(f(F)) \leq C_\alpha(F)$.

(3) Assume $\{x \in \bar{\bar{F}} : U_2^\gamma(x) < 1\} = \emptyset$. Then U_2^γ is continuous on the support of γ and thus everywhere: $U_2^\gamma(x)$ solves the exterior Dirichlet problem in G_∞ with boundary values 1.

(4) If F is a closed, sufficiently smooth surface, then the surface density of the measure γ is related to the normal derivative in the following way,

$$\lim_{x \to y} \frac{\partial U_2^\gamma}{\partial n}(x) = \frac{[\text{ surface density of } \gamma \text{ at } y]}{4\pi^2},$$

where n is the inward normal, see figure 8.4.

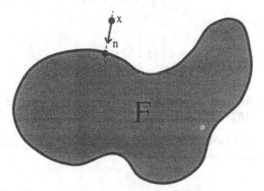

FIGURE 8.4. The surface density of the equilibrium measure on the boundary is connected to the normal derivative of the potential.

8.10. Logarithmic capacity. Let $N = 2$ and $\alpha = 2$, then the energy integral

$$I_2(\mu) = \int_F \int_F \log \frac{1}{|x - y|} \, d\mu(x) \, d\mu(y),$$

where we omit the factor $\omega_2 = 2\pi$.

If $F \subset \{|x| < 1\}$, we know that $I_2(\mu) \geq 0$. If $\max_{x \in F} |x| = R < 1$, then $I_2(\mu) \geq \mu(F)^2 \log \frac{1}{R} > 0$. Now, look at $\inf\{I_2(\mu) : \mu \in \overset{\circ}{\mathfrak{M}}{}_F^+\}$. We know that there exists a unique minimizing measure $\lambda = \lambda_F \in \overset{\circ}{\mathfrak{M}}{}_F^+$ such that $C_2(F) = I_2(F)^{-1}$. This is sometimes called "Wiener capacity".

There is also another variant.

DEFINITION 8.45 (LOGARITHMIC CAPACITY).

$$C_l(F) = \exp(-\frac{1}{C_2(F)}).$$

Since $0 \le C_2(F) < \infty$ we have $0 \le C_l(F) < 1$.

A useful property of C_l is the fact that

$$C_l(\{x : |x| < r\}) \approx r,$$

which is an analogue to the capacity of balls in higher dimensions.

However, there are drawbacks. In [10, p. 803] Doob says: " H. Jackson kindly showed me an example demonstrating that the logarithmic capacity set function is not subadditive."

EXERCISE 8.12. *Find such an example!*

On the other hand, $C_2(\cdot)$ is subadditive.

8.11. Polar sets. The polar sets are the exceptional sets in potential theory.

DEFINITION 8.46 (α–POLAR). $E \subset \mathbb{R}^N$ *is α-polar if there exists a measure μ such that*

$$(18) \qquad\qquad U_\alpha^\mu(x) = \begin{cases} +\infty & x \in E \\ < \infty & x \notin E. \end{cases}$$

E is a G_δ-set since we have,

$$E = \bigcap_{n=1}^{\infty} \{x : U_\alpha^\mu(x) > n\}.$$

THEOREM 8.47.   (1) *If $E \subset \mathbb{R}^N$ is α-polar, then $C_\alpha(E) = 0$.*

(2) *Suppose $E \subset \mathbb{R}^N$ is a G_δ-set and $C_\alpha(E) = 0$, then there is a measure λ such that formula (18) holds.*

DEFINITION 8.48 (GREENIAN SET). *An open subset of $\mathbb{R}^N$ supporting a positive non-constant superharmonic function.*

Example.

Let $N \ge 3$ and let $f(x) := |x|^{2-N}$. Then every open subset of $\mathbb{R}^N$ is Greenian.

THEOREM 8.49. *A non-empty open subset D of $\mathbb{R}^2$ is Greenian if and only if $\mathbb{R}^2 \setminus D$ is not polar.*

We used the notion of *polar* sets, which is slightly different from the above α–polar sets.

DEFINITION 8.50 (POLAR). *A set $E \subset \mathbb{R}^N$ is polar if to each point $x_0 \in E$, there exists a neighbourhood $\mathcal{O}$ of x_0 such that there is a superharmonic function in $\mathcal{O}$ which is $+\infty$ on $\mathcal{O} \cap E$.*

THEOREM 8.51. *Let $N \ge 2$. An open subset D of $\mathbb{R}^N$ is Greenian if and only if the Green function G_D exists.*

Doob defines a new energy integral in [10, chapter XIII] in the following way. Let $D \subset \mathbb{R}^N$ such that there is a Green function G_D. Define the energy integral

$$[\mu, \nu] := \int_D G_D \mu \, d\nu = \int_D \left(\int_D G_D(\xi, \eta) \mu \, d\eta \right) d\nu(\xi).$$

We have also $[\mu, \mu] \ge 0$ and we can "build" a capacity on this new energy integral.

8.12. A classical connection. Let $D \subset \mathbb{R}^N$ and let

$$\mathcal{D}(u,v) := \int_D \nabla u \cdot \nabla v \, dx.$$

The Dirichlet integral is then

$$\mathcal{D}(u,u) := \int_D |\nabla u|^2 \, dx.$$

If $u = G_D \mu$ and if $v = G_D \nu$ then $\Delta v = -c_N \nu$ and if we furthermore assume ∂D to be smooth,

$$[\mu, \nu] = \left(\int G_D \mu \, d\nu = \right) - c_N^{-1} \int_D u \Delta v = c_N^{-1} \mathcal{D}(u,v),$$

where we used Green's formula in the last step. For a discussion of these formulas, we refer to [24].

8.13. Another definition of capacity. From the above it is natural to define a new capacity,

$$cap(F) := \inf\{\mathcal{D}(u,u) : u \in C_0^\infty, u \geq 1 \text{ on } F\}.$$

This can be generalized. If $F \subset \mathbb{R}^N$ and $p \in (1, \infty)$ is given we define

$$cap_p(F) := \inf\{\int_{\mathbb{R}^N} |\nabla u|^p \, dx : u \in C_0^\infty, u \geq 1 \text{ on } F\}.$$

If $\mathcal{O}$ is an open set, we define

$$cap_p(\mathcal{O}) := \sup_{F \subset \mathcal{O}} cap_p(F), \quad \text{where } F \text{ is compact.}$$

If E is an arbitrary set, we define

$$cap_p(E) := \inf_{\mathcal{O} \supset E} cap_p(\mathcal{O}), \quad \text{where } \mathcal{O} \text{ is open.}$$

$cap_p(E)$ is called the (variational) p-capacity of E. As usual, there are now two definitions of the p-capacity of a compact set F: fortunately, they give the same number.

If $p = 2$, the formal discussion above shows that a minimizer of the energy integral is a solution of Laplace's equation $\Delta u = 0$ outside F. If $p \neq 2$ and $1 < p < \infty$, the corresponding differential equation is

$$\nabla \cdot (|\nabla u|^{p-2} \nabla u) = 0.$$

The corresponding operator is called the p–Laplace–operator: it is a non–linear operator and we are led to *nonlinear potential theory* . (cf. Heinonen–Kilpeläinen–Martio [25].)

9. Reduced functions

Let Ω be an open subset of $\mathbb{R}^N$ and suppose that Ω has a Green function. (cf. Definition 8.48 and Theorems 8.49, 8.51). Let $u \geq 0$ be a superharmonic function in Ω and let $E \subset \Omega$. The **reduced function** of u is defined as

$$R_u^E(x) := \inf\{v(x) : v \in \Phi_u^E\}, \text{where}$$

$$\Phi_u^E := \{v \text{ superharmonic in } \Omega, v \geq 0, v \geq u \text{ on } E\}.$$

Example 1. Let $\Omega = \{x_1 > 0\}$ and let E be a ball centered at $(2, 0, 0, \dots)$ with radius 1. Furthermore, let $u \equiv 1$ on E. Then $R_u^E \equiv 1$ on E and R_u^E is harmonic in $\Omega \setminus E$. The reduced function for the case $N = 2$ is depicted in Figure 9.1 below.

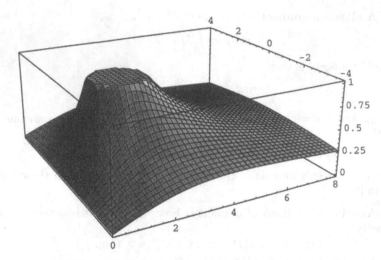

FIGURE 9.1. The reduced function in Example 1.

Example 2. Let $\Omega = \mathbb{R}^3$, $E = \{0\}$ and $u(x) = |x|^{-2}$. Then, $v(x) = \epsilon|x|^{-2} \in \Phi_u^E$ for every $\epsilon > 0$. This will give us that the reduced function is

$$R_u^E(x) = \begin{cases} +\infty & x = 0 \\ 0 & x \neq 0. \end{cases}$$

Note that this reduced function is *not* lower semi-continuous (l.s.c.). This can be dealt with by the following construction of the **regularized reduced function** $\hat{R}_u^E$.

$$\hat{R}_u^E(x) := \liminf_{y \to x} R_u^E(y).$$

We can now state some results about regularized reduced functions.

LEMMA 9.1. $\hat{R}_u^E$ *is superharmonic on* Ω.

PROOF. First of all we see that $0 \leq R_u^E \leq u$. What about the superharmonic mean value property? Let v be a function in Φ_u^E. Then

$$v(x) \geq \fint_{|y|=\rho} v(x+y)\,d\sigma_y \geq \fint_{|y|=\rho} R_u^E(x+y)\,d\sigma_y.$$

It follows that

$$R_u^E(x) \geq \fint_{|y|=\rho} R_u^E(x+y)\,d\sigma_y.$$

By Fatou's lemma we deduce

$$\hat{R}_u^E(x) \geq \fint_{|y|=\rho} \hat{R}_u^E(x+y)\,d\sigma_y.$$

Finally, we see that the set

$$\{x \in \Omega : \hat{R}_u^E(x) > a\} \text{ is open.}$$

Therefore $\hat{R}_u^E(x)$ is l.s.c. and we are done. $\square$

LEMMA 9.2. *The following properties of reduced functions hold.*

  (i) $u \geq R_u^E \geq \hat{R}_u^E \geq 0$ *on* Ω.
  (ii) $u = R_u^E$ *on* E.

(iii) $u = R_u^E = \hat{R}_u^E$ on the interior of E.

(iv) $R_u^E = \hat{R}_u^E$ on $\Omega \setminus \bar{E}$, and both functions are harmonic on $\Omega \setminus \bar{E}$.

LEMMA 9.3. Let $K \Subset \Omega$ (i.e. the closure of K is a compact subset of Ω). Then $\hat{R}_u^K(x)$ is a Green potential $G\mu$ in Ω.

For the proof see [26, Thm. 7.12].

THEOREM 9.4. (Cartan) Let $\mathfrak{F}$ be a family of superharmonic functions which is locally bounded below in Ω. Then $u = \inf\{v : v \in \mathfrak{F}\}$ differs from $\hat{u}$ at most on a set of capacity zero.

We immediately get the following Corollary.

COROLLARY 9.5. $\hat{R}_u^E(x) = u$ q.e. on E.

For the proof of the Theorem 9.4, we need the following lemma of Choquet which is stated and proved in [26] Lemma 2.22.

LEMMA 9.6. Let $\{f_i, i \in I\}$ be a family of functions on an open set Ω'. If $J \subset I$, let

$$f_J(x) = \inf_{i \in J} f_i(x), \quad x \in \Omega'.$$

Then there exists a countable set $I_0 \subset I$ which is such that if g is lower semi-continuous on Ω' and $g \leq f_{I_0}$ then $g \leq f_I$.

Proof of Theorem 9.4. Assume first that Ω has a Green function G and that all functions in $\mathfrak{F}$ are non-negative. Let B be an open ball such that $\bar{B} \subset \Omega$.

(1) $v = \hat{R}_v^B$ on B for all $v \in \mathfrak{F}$.

(2) $\hat{R}_v^B = G\mu$, $\operatorname{supp}(\mu) \subset \bar{B}$ for all $v \in \mathfrak{F}$.

(3) Apply Choquet's lemma with $\Omega' = B$ to $\{v|_B, v \in \mathfrak{F}\}$. Find $\{v_i\}_1^\infty$ such that $v_1 \geq v_2 \geq \ldots \forall x \in B$. Then we have,

$$\hat{R}_{v_i}^B = G\mu_i \text{ and } G\mu_1 \geq G\mu_2 \geq \ldots$$

(4) Find a measure σ such that $G\sigma = 1$ on B. This can be done by considering a slightly larger ball B', where $B \subset B' \subset \Omega$ and let $G\sigma := \hat{R}_1^{B'}$. Then we have

$$\mu_i(\bar{B}) = \int G\sigma \, d\mu_i = \int G\mu_i \, d\sigma \leq$$

$$\leq \int G\mu_1 \, d\sigma \leq \int G\sigma \, d\mu_1 = \mu_1(\bar{B}).$$

(5) Choose a weakly convergent subsequence (also called $\{\mu_i\}$) such that $\mu_i \rightharpoonup \mu$. Hence

$$G\mu \leq \liminf G\mu_i = \lim G\mu_i.$$

(6) Consider the Borel set

$$E = \{x \in B : G\mu(x) < \lim_{i \to \infty} G\mu_i(x)\}.$$

We now want to prove $c(E) = 0$, i.e. the Newtonian- (when $N \geq 3$) or the logarithmic capacity (in the plane case) of the set E is zero. To achieve that, assume that $c(E) > 0$. According to Theorem 8.39 there exists a compact subset F of E such that $c(F) > 0$. The proof of Theorem 4.3 tells us that there exists a nonzero measure ν with support in F and such that $G\nu$ is finite and continuous on $\bar{B}$. Thus we have

$$\int G\nu\, d\mu_i \to \int G\nu\, d\mu.$$

We conclude (by Fatou's lemma)

$$\int \lim G\mu_i\, d\nu \le \lim \int G\mu_i\, d\nu =$$

$$\lim \int G\nu\, d\mu_i = \int G\nu\, d\mu = \int G\mu\, d\nu.$$

Hence

$$\int (G\mu - \lim G\mu_i)\, d\nu \ge 0.$$

This is false due to the definition of the set E and the fact that $\text{supp}(\nu) \subset F \subset E$. Hence $c(E) = 0$.

(7) We know now that

$$G\mu(x) \le \inf\{v_i : i = 1, 2, \dots\}.$$

Thus, due to the lemma of Choquet

$$G\mu(x) \le \inf\{v|_B :\in \mathfrak{F}\}.$$

Therefore

$$G\mu(x) \le u(x) \text{ in } B.$$

Hence

$$G\mu(x) \le \hat{u}(x) \le u(x) \le \lim_{i \to \infty} G\mu_i(x) \text{ in } B.$$

This chain of inequalities leads to

$$\{x \in B : u(x) > \hat{u}(x)\} \subset \{x \in B : G\mu(x) < \lim_{i \to \infty} G\mu_i(x)\}$$

and

$$c(\{x \in B : u(x) > \hat{u}(x)\}) \le c(\{x \in B : G\mu(x) < \lim_{i \to \infty} G\mu_i(x)\}) =$$

$$= c(E) = 0.$$

Hence

$$c(\{x \in B : u(x) > \hat{u}(x)\}) = 0.$$

Covering Ω with countably many balls, we conclude that

$$c(\{x \in \Omega : u(x) > \hat{u}(x)\}) = 0.$$

Let us now study the general case when we do not assume that Ω has a Green function. Choose $\Omega_1 \Subset \Omega$ such that Ω_1 has a Green function. Let

$$\alpha = \inf\{v(x) : v \in \mathfrak{F}, \ x \in \Omega_1\} > -\infty.$$

Then, let $w = v - \alpha \ge 0$ for all $v \in \mathfrak{F}$. We can now apply the previous case to $\{w = v - \alpha, \ v \in \mathfrak{F}\}$ in Ω_1. Let $W := \inf\{w_1, w_2, \dots\}$, then we have $W - \hat{W} = u - \hat{u}$ and

$$c(\{x \in \Omega_1 : u(x) > \hat{u}(x)\}) = 0.$$

We can now repeat this for a denumerable sequence $\{\Omega_i\}_1^\infty$ such that

$$\Omega = \bigcup_{\Omega_i \subset\subset \Omega} \Omega_i.$$

This ends the proof of Theorem 9.4.

In [24, chapters 7 and 8], Hedberg discusses *balayage* in the case $\Omega = \mathbb{R}^N$. Balayage and regularized reduced functions are strongly related. If F is compact we claim that

$$U^{\mu_F} = \hat{R}_u^F \text{ if } u = U^\mu, \text{ where } U^{\mu_F} \text{ is the balayage of } U^\mu \text{ in } \mathbb{R}^N.$$

Recall that

- $U^{\mu_F} = U^\mu = u$ q.e. on F.
- $U^{\mu_F} \leq U^\mu$ everywhere.
- U^{μ_F} is harmonic in $\mathbb{R}^N \setminus F$.
- The support of μ_F is a subset of F.

If we let $v \in \Phi_u^F$ then $v \geq U^{\mu_F}$. Thus $\hat{R}_u^F \geq U^{\mu_F}$ and

$$u \stackrel{\text{q.e. on } F}{=} U^{\mu_F} \leq \hat{R}_u^F \leq R_u^F \stackrel{\text{on } F}{=} u.$$

Hence we end up with two potentials where the following holds.

(i) The Riesz masses are both subsets of F.
(ii) $\hat{R}_u^F = U^{\mu_F}$ q.e. on F.
(iii) $\hat{R}_u^F \geq U^{\mu_F}$ everywhere.
(iv) Both potentials are harmonic in $\mathbb{R}^N \setminus F$.
(v) Both potentials are 0 at ∞.

If both potentials are uniformly bounded we must have $\hat{R}_u^F = U^{\mu_F}$ in $\mathbb{R}^N \setminus F$ and then (ii) and (iv) would give us that $\hat{R}_u^F = U^{\mu_F}$ everywhere.

To handle the case when u is unbounded, we consider

$$u_n := \min(u, n) =: U^{\mu_n}.$$

We know that $\hat{R}_{u_n}^F = U^{\mu_n F}$. According to [24, Lemma 8.2], we have $U^{\mu_n F} \nearrow U^{\mu_F}$ everywhere as $n \to \infty$.

We claim that $\hat{R}_{u_n}^F \nearrow \hat{R}_u^F$ as $n \to \infty$. To prove this, we note first that

$$R_{u_n}^F = u_n \nearrow u = R_u^F \quad \text{on } F$$

(cf. Lemma 9.2). Applying Theorem 9.4, we see that $\hat{R}_{u_n}^F \nearrow \hat{R}_u^F$ q.e. on F as $n \to \infty$. Furthermore, $\hat{R}_{u_n}^F$ increases to a superharmonic function majorized by the potential U^{μ_F}. According to Corollary 8.22, the limit is a Green potential $G\nu$ in $\mathbb{R}^N$. We conclude that $G\nu = \hat{R}_u^F$ q.e. on F and that $G\nu \leq \hat{R}_u^F$ everywhere.

We note that $\hat{R}_u^F = G\nu'$ is also a Green potential on $\mathbb{R}^N$. Since

$$\int G\nu \, d\lambda = \int G\nu' \, d\lambda$$

for all λ with finite energy and support in F, it follows from [24, Corollary 8.6] that $\nu_F = \nu'_F$. Since $\text{supp}\nu$ and $\text{supp}\nu'$ are contained in F, we conclude that $\nu = \nu_F = \nu'_F = \nu'$ which proves that $\hat{R}_u^F = \lim \hat{R}_{u_n}^F = U^{\mu_F}$.

10. Green energy in a Half–space

Let $D = \{x \in \mathbb{R}^N : x_1 > 0\}$. Furthermore, let $\tilde{y} = (-y_1, y_2, \dots)$ be the reflection of y in ∂D.

DEFINITION 10.1 (GREEN FUNCTION).

$$G(x, y) = \begin{cases} |x - y|^{2-N} - |x - \tilde{y}|^{2-N} & \text{if } N \geq 3 \\ \log(|x - \tilde{y}|/|x - y|) & \text{if } N = 2. \end{cases}$$

We also have the Poisson kernel,

$$P(x, y) := c_N x_1 |x - y|^{-N}, \quad N \geq 2,$$

where

$$P(x, y') = \lim_{y_1 \searrow 0} G(x, y)/y_1, \quad y = (y_1, y'), \quad y' \in \partial D.$$

We see that G is defined on $D \times D$ and P on $D \times \partial D$. Furthermore, we can define the **Green potential** $G\mu(x)$ and the **Poisson integral** $P\nu(x)$ as

$$G\mu(x) := \int_D G(x, y) \, d\mu(y),$$

$$P\nu(x) := \int_{\partial D} P(x, y) \, d\nu(y).$$

Let now E be a subset of the half-space D. Suppose there exists a measure λ on D such that $\hat{R}^E_{x_1} = G\lambda_E$. The measure λ_E is called **the fundamental distribution** on E. We know that if E is compact there exists such a measure with support in E.

DEFINITION 10.2. *The Green energy $\gamma(E)$ is defined by*

$$\gamma(E) = \int (G\lambda_E) \, d\lambda_E.$$

We shall call $\lambda_E(D)$ the outer charge of D. The concept of Green energy was introduced by J. Lelong-Ferrand in 1949. See [30]. She did not use the name "Green energy" but "puissance extérieure" which can be translated as *outer charge*: her *outer charge* is different from our *outer charge*.

10.1. Properties of the Green energy γ. The following lemma can be found in [13, Lemma 2.1].

LEMMA 10.3. *γ is a countably sub–additive set function such that*

(1) *$\gamma(E_n) \nearrow \gamma(E)$ if $E_n \nearrow E$.*
(2) *$\gamma(E) = \inf\{\gamma(\mathcal{O}) : \mathcal{O} \supset E, \mathcal{O} \text{ open }\}$.*

Hence γ is a capacity (c.f. Section 2.2).
Alternatively, we can introduce a set–function γ' as follows. We argue as in Section 5, replacing the class Γ_E by

$$\Gamma'_E := \{\mu : \mu \text{ pos. measure, supp}(\mu) \subset E \text{ and } G\mu(x) \leq x_1 \text{ q.e. on } E\}.$$

Define $\gamma'(E) := \sup_{\mu \in \Gamma'_E} \int x_1 \, d\mu(x)$. If F is compact, $\hat{R}^F_{x_1} = G\lambda_F$ and if furthermore $\mu \in \Gamma'_F$ then

$$\int x_1 \, d\mu = \int G\lambda_F \, d\mu = \int G\mu \, d\lambda_F \leq \int x_1 d\lambda_F = \gamma(F).$$

Hence $\gamma'(F) \leq \gamma(F)$. Since $\lambda_F \in \Gamma'_F$ we conclude that

$$\gamma'(F) = \gamma(F).$$

Let us now state an analogue of Theorem 5.7.

THEOREM 10.4. *Let F be a compact subset in the half–space D and let*

$$B = \inf\{\int x_1 \, d\mu, G\mu(x) \geq x_1 \quad q.e. \text{ on } F\}.$$

Then $B = \gamma(F)$.

Remark. The fundamental distribution λ_E plays the same role for γ as the previous equilibrium distribution plays in the theory of Newtonian capacity.

EXERCISE 10.1. *Prove that γ is subadditive!*

10.2. Ordinary thinness.

DEFINITION 10.5 (FINE TOPOLOGY). *[[26], Chapter 10]. The fine topology on $\mathbb{R}^N$ is the smallest topology on $\mathbb{R}^N$ for which all superharmonic functions are continuous in the extended sense.*

Remark. The class of sets of the forms

$$\{y = v(y) < \alpha\}, \{y = v(y) > \beta\},$$

where α, β are arbitrary real numbers and v an arbitrary superharmonic function, forms a basis for this topology. $\{y = v(y) > \beta\}$ is open in the metric topology and $\{y = v(y) < \alpha\}$ gives more neighbourhoods. The fine topology contains more open sets than the metric topology, i.e. it is "finer". We know that there exist discontinuous superharmonic functions.

DEFINITION 10.6. *A set E is **thin** at x if x is not a fine limit point of E.*

Let us now state some results quoted from [26].

THEOREM 10.7. [26, Theorem 10.3].
A set E is thin at a limit point of E if and only if there is a superharmonic function u on a neighbourhood of x such that

$$u(x) < \liminf u(y), \ y \to x, \ y \in E \setminus \{x\}.$$

THEOREM 10.8. [26, Theorem 10.4].
A set E is thin at a limit point x of E if and only if there is a superharmonic function u on a neighbourhood of x such that

$$u(x) < \lim_{y \to x, \, y \in E \setminus \{x\}} u(y) = \infty.$$

DEFINITION 10.9. *An extended real-valued function u defined on an open set Ω **peaks** at x if*

$$u(x) > \sup u(y), \ y \in \Omega \setminus V$$

for every neighbourhood V of x.

If Ω has a Green function G, $G(x, \cdot)$ peaks at x.

THEOREM 10.10. [26, Theorem 7.7]. *Let Ω be an open set having a Green function and let u be a positive superharmonic function on Ω which peaks at x. A set $E \subset \Omega$ is thin at x if $\hat{R}_u^E(x) < u(x)$.*

THEOREM 10.11. [26, Theorem 10.11].
Let F be closed subset of $\mathbb{R}^N$. Then F is not thin at $x \in \partial F$ if and only if there exists a positive superharmonic function w defined on $\complement F$ such that

$$0 = \lim_{y \to x, \, y \notin F} w(y).$$

In [24, Proposition 9.10], we find a necessary and sufficient condition for a set to be thin at a limit point: it is sometimes used as a definition of thinness.

THEOREM 10.12. *A set F is **thin** at a limit point a of F if and only if there exists a measure $\mu \in \mathcal{M}^+$ such that*

$$U^\mu(a) < \liminf_{x \to a,\, x \in F \setminus \{a\}} U^\mu(x).$$

11. Minimal thinness

A general definition is given in [10, 1.XII.11]. Let Ω be a ball $B \subset \mathbb{R}^N$ or a half space $D = \{x \in \mathbb{R}^N, x_1 > 0\}$. If y is a boundary point, we define h to be the Poisson kernel at y if y is finite or $h(x) = x_1$ if $y = \infty$. We say that $E \subset \Omega$ is **minimally thin** at y if there exists a point z in Ω such that $\hat{R}_h^E(z) < h(z)$. (The reduced function is formed with respect to all non-negative superharmonic non-negative functions in Ω.)

We shall discuss minimal thinness at infinity, i.e. the case $y = \infty$, and $h(x) = x_1$. If $\Omega \setminus E$ is open and $\hat{R}_h^E(z) < z_1$ for a point z in Ω then $\hat{R}_h^E(x) < x_1$ for all $x \in \Omega \setminus E$ due to the maximum principle.

11.1. Minimal thinness, Green potentials and Poisson integrals. If u is a superharmonic non-negative function in D then the Riesz representation theorem [26, Theorem 6.18] says that

$$(19) \qquad u(x) = \alpha x_1 + G\mu(x) + P\nu(x),$$

where $\alpha \geq 0$ and μ, ν are non-negative measures.

$G\mu$ is defined if and only if

$$(20) \qquad \int y_1 (1 + |y|)^{-N} \, d\mu(y) < \infty.$$

We shall need the following estimates.

$$(21) \qquad c_N x_1 y_1 |x - \tilde{y}|^{-N} \leq G(x, y) \leq c_N' x_1 y_1 |x - y|^{-N}, \quad (N \geq 2),$$

$$(22) \qquad G(x, y) \leq c_N'' x_1 y_1 |x - y|^{2-N} |x - \tilde{y}|^{-2}, \quad (N \geq 3),$$

$$(23) \qquad G(x, y) \leq |x - y|^{2-N}, \quad (N \geq 3).$$

From the convergence of the integral in (20), it follows that

$$(24) \qquad \lim_{R \to \infty} R^{-N} \int_{|y| \leq R} y_1 \, d\mu(y) = 0.$$

We can state the following lemma

LEMMA 11.1. *Let $E = \{x \in D,\ G\mu(x) \geq x_1\}$. Then E is minimally thin at ∞ for any measure μ for which $G\mu$ is defined.*

PROOF. Let

$$v(x) = v_R(x) = \begin{cases} x_1 & \text{as } |x| \leq R, x \in D \\ x_1 R^N |x|^{-N} & \text{as } |x| > R, x \in D. \end{cases}$$

Note that $v(x)$ is superharmonic in D and, since $v = 0$ on ∂D, that $v = G\lambda_R$, where λ_R is the Riesz mass of v (c.f. the example after Corollary 8.30). The Riesz mass of v is $-C_N \Delta v$ (the Laplacian is taken in the sense of distributions.) By the use of

Green's formula where $\phi \in C_0^\infty(D)$ is a test function, $D_R := D \cap \{|x| < R\}$ and n_- the inner unit normal and n_+ the outer unit normal to D_R (see figure 11.1)

$$(25) \qquad \int_{D_R} v \Delta \phi \, dx = \int_{D_R} (v \Delta \phi - \phi \Delta v) \, dx = - \int_{|x|=R} (v \frac{\partial \phi}{\partial n_-} - \phi \frac{\partial v}{\partial n_-}) \, d\sigma,$$

$$(26) \qquad \int_{D \backslash D_R} v \Delta \phi \, dx = \int_{D \backslash D_R} (v \Delta \phi - \phi \Delta v) \, dx = - \int_{|x|=R} (v \frac{\partial \phi}{\partial n_+} - \phi \frac{\partial v}{\partial n_+}) \, d\sigma.$$

In (25) we have that

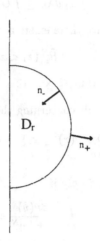

FIGURE 11.1. The unit normals to D_r.

$$\frac{\partial v}{\partial n_-} = -\frac{\partial}{\partial r}(r \cos \theta) = -\cos \theta$$

and in (26)

$$\frac{\partial v}{\partial n_+} = \frac{\partial}{\partial r}(R^N r^{1-N} \cos \theta)|_{r=R} = (1 - N) \cos \theta.$$

Since $\nabla \phi$ is continuous

$$\frac{\partial \phi}{\partial n_-} + \frac{\partial \phi}{\partial n_+} = 0 \text{ on } \{|x| = R\}.$$

Hence we have

$$\int_D v \Delta \phi = \int_{|x|=R} \phi(\frac{\partial v}{\partial n_+} + \frac{\partial v}{\partial n_-}) d\sigma_R =$$

$$= -N \int \phi \cos \theta \, d\sigma_R(x) = -N \int \phi \frac{x_1}{R} \, d\sigma_R(x).$$

We conclude that

$$d\lambda_R(x) = C_N(x_1/R) \, d\sigma_R(x).$$

We will then have

$$\int G\mu \, d\lambda_R = \int G\lambda_R \, d\mu = \int_{|x| \le R} x_1 \, d\mu(x) + R^N \int_{|x| > R} x_1 |x|^{-N} \, d\mu(x).$$

By (24) it follows that

$$\int G\mu \, d\lambda_R = \underline{o}(R^N) \text{ as } R \to \infty.$$

We have also,

$$\int x_1 \, d\lambda_R = c_N R^N,$$

where $c_N > 0$. If E is not minimally thin at ∞ then

$$x_1 = \hat{R}_{x_1}^E \le G\mu(x),$$

which leads us to

$$c_N R^N = \int x_1 \, d\lambda_R \le \int G\mu \, d\lambda_R = \varrho(R^N),$$

which is a contradiction. Hence there exists a $z \in D$ such that

$$\hat{R}_{x_1}^E(z) < z_1$$

and therefore E is minimally thin at ∞. $\square$

There is also an analogous result concerning the Poisson integral.

LEMMA 11.2. $E = \{x \in D, P\nu(x) \ge x_1\}$ *is minimally thin at ∞ for any measure on ∂D for which $P\nu$ is defined.*

PROOF. $P\nu$ is defined if and only if

$$\int_{\partial D} \frac{d\nu(y)}{1 + |y|^N} < \infty,$$

which implies

$$\lim_{R \to \infty} R^{-N} \nu(\partial D \cap \{|y| \le R\}) = 0.$$

Now, if $y \in \partial D$ and $y = (0, y_2, \dots)$ and $y_t = (t, y_2, \dots)$, then we have

$$P(x, y) = \lim_{t \searrow 0} \frac{G(x, y_t)}{t} = \frac{\partial G}{\partial n}(x, y).$$

We have also

$$\int P\nu \, d\lambda_R = c_N \int d\nu(y) \int P(x, y) \, d\lambda_R(x).$$

Since

$$\int G(x, y) \, d\lambda_R(x) = v_R(y),$$

it is clear that

$$\int P(x, y) \, d\lambda_R(x) \le \liminf_{t \searrow 0} \int \frac{G(x, y_t)}{t} \, d\lambda_R(x) =$$

$$= \liminf_{t \searrow 0} \frac{v(y_t)}{t} = \begin{cases} 1 & \text{as } |y_t| \le R, \, y_t \in D \\ R^N |y|^{-N} & \text{as } |y_t| > R, \, y_t \in D. \end{cases}$$

Therefore

$$\int P\nu \, d\lambda_R \le c_N \nu(\{|y| \le R\}) + R^N \int_{|y| \ge R} \frac{d\nu(y)}{|y|^N} = \varrho(R^N)$$

By using the same argument as in the proof of Lemma 11.1 we obtain Lemma 11.2. $\square$

If E is minimally thin at ∞, $\hat{R}^E_{x_1}$ is a superharmonic function in D. Riesz theorem then implies

$$\hat{R}^E_{x_1} = \alpha x_1 + G\mu(x).$$

Since $\hat{R}^E_{x_1} \leq x_1$ we must have $\alpha < 1$. Note that on E

$$G\mu(x) = (1-\alpha)x_1 \quad \text{q. e.}.$$

We can now normalize by putting $\mu_1 := \frac{\mu}{1-\alpha}$.
Thus, we have for E found a measure μ_1 such that

$$E \subset \{x \in D : G\mu_1(x) \geq x_1\}.$$

11.2. A criterion of Wiener type for minimal thinness. Let $I_n = \{x \in D : 2^n \leq |x| < 2^{n+1}\}, n = 1, 2, \ldots$ and let $E_n = E \cap I_n$.

THEOREM 11.3. *(J. Lelong-Ferrand, [30])*
E is minimally thin at infinity in D if and only if

$$\sum \gamma(E_n) 2^{-nN} < \infty.$$

PROOF. Assume that E is minimally thin at infinity. Then there exists a measure μ such that the Green potential $G\mu$ has the property

$$E \subset \{x \in D : G\mu(x) \geq x_1\}.$$

(i) Let μ_0 be the restriction of μ to $D \cap \{|x| \leq R\}$ where R is chosen so that for $|x| \geq 2R$

$$G\mu_0(x) \leq \text{Const.}x_1 \int y_1 |x-y|^{-N} d\mu_0(y) \leq$$

$$\leq \text{Const.}x_1 R^{-N} \int_{|y| \leq R} y_1\, d\mu(y) \leq x_1/4.$$

Furthermore, we can assume that if $|x| \geq 2R$,

(27) $$\text{Const.} \int_D y_1 \inf\{|x|^{-N}, |y|^{-N}\}\, d\mu(y) \leq 1/4$$

If $\mu_1 := \mu - \mu_0$, it follows that

$$E \cap \{|x| \geq 2R\} \subset \{x \in D : G\mu_1(x) \geq 3x_1/4\}$$

(ii) If $x \in I_n$ and $|x| \geq 4R$, we write

$$L_n := I_{n-1} \cup I_n \cup I_{n+1},$$

and obtain

$$G\mu_1(x) = \{\int_{L_n} + \int_{D \backslash L_n}\} G(x,y)\, d\mu_1(y) = (A_1 + A_2)(x).$$

Then

$$A_2(x) \leq \text{Const.}\{x_1 |x|^{-N} \int_{|y| \leq |x|/2} y_1\, d\mu_1(y) +$$

$$+ x_1 \int_{|y| \geq 2|x|} y_1 |y|^{-N} d\mu_1(y)\} \leq x_1/4 \quad \text{(c.f. equation (27).)}$$

Hence

$$E_n = E \cap I_n \subset \{x \in I_n : A_1(x) \geq x_1/2\}.$$

(iii) Let λ_n be the fundamental distribution on E_n. Then $G\lambda_n(x) = x_1$ q.e. on E_n giving

$$\gamma(E_n) = \int G\lambda_n \, d\lambda_n \leq 2 \int \{\int_{L_n} G(x,y) \, d\mu(y)\} \, d\lambda_n(x) \leq 2 \int_{L_n} y_1 \, d\mu(y).$$

Since $G\mu$ exists, $\int y_1(1 + |y|)^{-N} \, d\mu(y) < \infty$ and it follows that

(28) $$\sum \gamma(E_n) 2^{-nN} < \infty.$$

Conversely, assuming (28), we let λ_n be the fundamental distribution on $E_n, n = 1, 2, \ldots$, and form

$$G\mu = \sum \int_{E_n} G(x,y) \, d\lambda_n(y)$$

Since we have (28), $G\mu(x)$ will be defined. Furthermore, we have

$$G\mu(x) \geq x_1 \quad \text{q.e. on } E = \cup E_n,$$

giving us that

$$E \subset \{x \in D : G\mu(x) \geq x_1\}.$$

Hence E is minimally thin and the theorem is proved. $\square$

THEOREM 11.4. *Let u be a nonnegative superharmonic function in D. Then there exists a set $E \subset D$ which is minimally thin at infinity in D and a constant α such that*

(29) $$\lim u(x)/x_1 = \alpha, x \to \infty, x \in D \setminus E.$$

Conversely, to each set E which is minimally thin at infinity in D, there exists a Green potential $u = G\mu$ such that $u(x) \geq x_1$ on E.

PROOF. By the Riesz representation theorem, u can be written as in (19), i.e. $u(x) = \alpha x_1 + G\mu(x) + P\nu(x)$. According to Lemmas 11.1 and 11.2, the set

$$E(\epsilon) = \{x \in D : G\mu(x) + P\nu(x) \geq \epsilon x_1\}$$

is minimally thin at infinity for each $\epsilon > 0$. Let $\{\epsilon_n\}$ decrease to 0 and choose $\{q(n)\}_{n=1}^{\infty}$ increasing to infinity such that

$$\sum_{n=1}^{\infty} \sum_{k=q(n)}^{\infty} \gamma(E(\epsilon_n) \cap I_k) 2^{-kN} < \infty.$$

From Theorem 11.3, we see that

$$E = \bigcup_{n=1}^{\infty} \bigcup_{k=q(n)}^{\infty} (E(\epsilon_n) \cap I_k)$$

is minimally thin at infinity. Now, if we take $x \in D \setminus E$ and $x \in I_m$ then

$$x \notin E(\epsilon_n) \cap I_m,$$

with $q(n) \leq m$. Thus

$$G\mu(x) + P\nu(x) < \epsilon_{q^{-1}(m)} x_1.$$

We see that (29) holds with E chosen in this way. For the proof of the converse statement, we refer to the proof of Theorem 11.3. $\square$

12. Rarefiedness

In Section 11, we compared the growth at infinity of a nonnegative superharmonic function u in D with the growth of the function x_1. Can we give a Wiener–type criterion for sets $\{x \in D : u(x) > |x|\}$? To solve this problem, we introduce a new capacity called the Green mass (cf. [13, Definition 2.3]) and a Martin type kernel on $\bar{D} \cup \{\infty\}$ as follows

$$K(y, x) = \begin{cases} G(y, x)/y_1, & x \in D, \ y \in D, \\ P(x, y), & x \in D, \ y \in \partial D, \\ x_1, & x \in D, \ y = \infty. \end{cases}$$

We note that $K(\cdot, x)$ is continuous on $\bar{D} \setminus \{x\}$.

DEFINITION 12.1. *Let $E \subset D$ be a bounded set. Then $\hat{R}_1^E(x)$ is a nonnegative superharmonic function in D and we have*

$$\hat{R}_1^E(x) = \int_{\partial D} K(y, x) \, d\mu_1(y) + \int_D G(y, x) \, d\mu_2(y)$$

where the measures μ_1 and μ_2 are nonnegative. Let λ'_E be the measure defined by

$$d\lambda'_E = \begin{cases} d\mu_1 & \text{on } \partial D \\ y_1 \, d\mu_2 & \text{on } D. \end{cases}$$

*We define the **Green mass** to be*

$$\lambda'(E) = \lambda'_E(E) := \mu_1(\partial D) + \int_D y_1 \, d\mu_2(y).$$

We quote Lemmas 2.4 and 2.5 from [13].

LEMMA 12.2. *λ'_E is a countably sub–additive set function such that*

(i) $\qquad\qquad \lambda'(E_n) \nearrow \lambda'(E)$ *if $E_n \nearrow E$,*

(ii) $\quad \lambda'_E(E) = \inf\{\lambda'(\mathcal{O}) : \mathcal{O} \supset E, \mathcal{O} \text{ open }\}.$

Hence $\lambda'_E(\cdot)$ is a capacity (cf. Section 2.2).

LEMMA 12.3. *The outer charge $\lambda_E(D)$ and the Green mass $\lambda'_E(E)$ coincide for any set $E \subset D$.*

To explain Lemma 12.3, we give an outline of the proof. Assume $\hat{R}_{x_1}^E$ is a potential $G\lambda_E$. If μ is a mass distribution on $\bar{D}$, we define

$$K\mu(x) := \int K(y, x) \, d\mu(y),$$

$$K^*\mu(x) := \int K(x, y) \, d\mu(y).$$

We note that $\hat{R}_1^E = K\lambda'_E$. Then we have

$$\lambda_E(D) = \int \hat{R}_1^E \, d\lambda_E = \int K\lambda'_E \, d\lambda_E = \int_{\bar{D}} K^*\lambda_E \, d\lambda'_E,$$

$$\lambda'_E(\bar{D}) = \int_D \frac{1}{y_1} \hat{R}_{x_1}^E(y) \, d\lambda'_E(y) + \lambda'_E(\partial D) = \int_D K^*\lambda_E \, d\lambda'_E + \lambda'_E(\partial D).$$

Let B_E be the set of points of ∂D where E is not minimally thin. Then according to [13, Lemma 2.3],

$$K^*\lambda_E = 1 \text{ on } B_E \text{ and } \lambda'_E(\partial D \setminus B_E) = 0.$$

Hence
$$\int_{\partial D} K^* \lambda_E \, d\lambda'_E = \int_{\partial D} 1 \, d\lambda'_E = \lambda'_E(\partial D),$$
which proves that $\lambda_E(D) = \lambda'_E(\bar{D})$.

DEFINITION 12.4. *A set E is rarefied at infinity in D if there exists a positive superharmonic function u in D with no Riesz mass at infinity such that*
$$u(x) \geq |x|, x \in E.$$

Remark. In Definition 3.2 in [13], rarefiedness was defined via a condition of Wiener type (cf. Theorem 13.1 below). The property used in the definition above is there given in Theorem 4.2.

Remark. In the definition of R_r^E, we assumed that u is superharmonic. The function $|x|$ is however subharmonic. If $r = |x|$, it is clear that R_u^E can be defined in the same way as R_u^E. We lose the analogue of Lemma 9.2 (ii). If E is rarefied at infinity in D, we see that $\hat{R}_r^E$ is defined and that $\hat{R}_r^E = G\mu_0 + P\nu_0$ where ν_0 and μ_0 are non-negative measures.

LEMMA 12.5. *The set $\{x \in D : G\mu(x) + P\nu(x) \geq |x|\}$ is rarefied at infinity for any measures μ and ν for which $G\mu$ and $P\nu$ are defined.*

PROOF. This is clear from the definition since any positive superharmonic function u in D with no Riesz mass at infinity can be written as
$$u(x) = G\mu(x) + P\nu(x).$$

$\square$

13. A criterion of Wiener type for rarefiedness

We will use the notation from Section 11.2.

THEOREM 13.1. *E is rarefied at infinity in D if and only if*

(30)
$$\sum \lambda'(E_n) 2^{-n(N-1)} < \infty.$$

We start the proof with the following lemma.

LEMMA 13.2. *Let F be relatively compact subset of $\bar{D}$ and assume that*
$$u_0 = G\mu_0 + P\nu_0$$
is such that $u_0(x) \geq 1$ q.e. on F. Then
$$\lambda'(F) \leq \int (y_1 \, d\mu_0(y) + d\nu_0(y)).$$

PROOF. According to Lemma 12.3, we know that $\lambda'(F) = \lambda(F)$ where $\lambda(F)$ is the outer charge of F, i.e. $\hat{R}_{x_1}^F = G\lambda$. Hence
$$\lambda'(F) = \lambda(F) \leq \int u_0(x) \, d\lambda_F(x) \leq$$
$$\leq \int_D y_1 \, d\mu_0(y) + \int_{\partial D} d\nu_0(y) \int P(x, y) \, d\lambda_F(x).$$

To estimate the second integral, we note that if $y = (0, y') \in \partial D$ and $y_t = (t, y') \in D$ then Fatou's lemma implies that
$$\int P(x, y) \, d\lambda_F(x) \leq \liminf_{t \searrow 0} \int \frac{1}{t} G(x, y_t) \, d\lambda_F(x) \leq 1.$$

The lemma follows. $\square$

PROOF OF THEOREM 13.1. Assume that E is rarefied at infinity in D. Hence there exists measures μ and ν such that.

$$E \subset \{x \in \bar{D} : u(x) = G\mu(x) + P\nu(x) \geq |x|\}.$$

Now, let $L_n := I_{n-1} \cup I_n \cup I_{n+1}$. If $x \in I_n$ and n is large enough, we argue as in Section 11.2 and see that

$$\{x \in I_n : u(x) \geq |x|\} \subset \{x \in I_n : A_1(x) \geq |x|/4\} =: E_n'$$

where

$$A_1(x) = \int_{\bar{L}_n} (G(x,y)\, d\mu(y) + P(x,y)\, d\nu(y)).$$

If $A_2(x) := u(x) - A_1(x)$ and $x \in I_n$ is large, then $A_2(x) \leq |x|/4$. When estimating u in I_n, it suffices to consider the contribution to the potential from $\bar{L}_n$.

Since $A_1(x) \geq C2^n$ for $x \in E_n'$, we apply the lemma with $u_0 = A_1/(C2^n)$ and conclude that

$$\lambda'(E_n') \leq C_1 2^{-n} \int_{L_n} (y_1\, d\mu(y) + d\nu(y)).$$

Summing, we obtain

$$\sum 2^{-n(N-1)} \lambda'(E_n') \leq C_1 \sum 2^{-Nn} \int_{L_n} (y_1\, d\mu(y) + d\nu(y)) < \infty.$$

The sum is convergent since we have assumed that u is defined. We conclude that (30) holds.

Conversely, assuming that (30) holds, we let λ_n' be such that $\hat{R}_1^{E_n} = K\lambda_n'$. Then $\lambda'(E_n) = \lambda_n'(\bar{D})$ and we define

$$u(x) := \sum 2^{n+1} K\lambda_n'(x).$$

Since (30) holds, u is defined.

Furthermore, for all n,

$$u(x) \geq 2^{n+1} \geq |x|, \quad \text{q.e. for } x \in E_n.$$

Thus $u(x) \geq |x|$ q.e. on $E = \cup E_n$ and the theorem is proved. $\square$

14. Singular integrals and potential theory

Let $D = \{x \in \mathbb{R}^N : x_1 > 0\}$. As in Section 10 let $G(x,y)$ be Green's function for D and $P(x,y)$ the Poisson kernel, $G\mu$ the Green potential of μ (where μ is supported on $\bar{D}$) and let $P\nu$ be the Poisson integral of ν, where ν is supported on ∂D. Changing the notation in the previous section, we define

$$K(x,y) = \begin{cases} G(y,x)/x_1 y_1 & x \in D, \quad y \in D, \\ P(x,y)/x_1 & x \in D, \quad y \in \partial D, \\ 0 & x \notin \bar{D} \text{ or } y \notin \bar{D}, \end{cases}$$

$$K\mu(x) := \int K(x,y)\, d\mu(y).$$

We note that $K(x,y)$ is continuous in $D \times \bar{D} \setminus \{x = y\}$. Next, we turn our interest to the following sets.

$$\{x \in D : G\mu(x) \geq x_1\} = \{x \in D : K\mu'(x) \geq 1\},$$

where $d\mu'(x) := x_1 \, d\mu(y)$ and

$$\{x \in D : P\nu(x) \geq x_1\} = \{x \in D : K\nu(x) \geq 1\}.$$

Following [2] we prove.

THEOREM 14.1. *Let* $f \in L^p(D)$ *and let* μ *be a measure with support in* $\bar{D}$. *Then the following holds.*

A) *Let* $1 < p < \infty$. *Then* $\|Kf\|_p \leq C_p\|f\|_p$.
B) *If* $\lambda > 0$, *then* $|\{x \in D : |K\mu(x)| \geq \lambda\}| \leq C\|\mu\|/\lambda$.
C) *If* E *is a measurable subset of* D *then* $|E| \leq C\gamma(E)$ *(where* $\gamma(E)$ *is the Green energy).*

COROLLARY 14.2.

$$|\{x \in D : G\mu(x) \geq x_1\}| \leq C \int x_1 \, d\mu(x),$$

$$|\{x \in D : P\nu(x) \geq x_1\}| \leq C\|\nu\|.$$

For the history of the above problem see [2, pp. 27, 29].

Remark 1. It is easy to see that $(B) \Rightarrow (C)$.

Remark 2. This method is applicable to $C^{1,\alpha}$–domains in D (we use estimates of Widman, cf. [2, p. 28]).

In the proof, we consider only functions or measures supported by $\bar{D}$. Let

$$Mf(x) := \sup_{x \in Q} |Q|^{-1} \int_Q |f| \, dx,$$

where the supremum is taken over all cubes in $\mathbb{R}^N$ with sides parallel to the coordinate axis. All cubes in this section will be of this type.

We prove first that (B) is true using that (A) holds for $p = 2$. It suffices to prove (B) assuming that μ is absolutely continuous and nonnegative, where $d\mu := f \, dx, f \geq 0$, and f has compact support in $\bar{D}$.

Following [33, I.3, II.2] we define $F := \{x : Mf(x) \leq \alpha\}$ and $\Omega := \{x : Mf(x) > \alpha\}$. We note that F is closed. Let $\{Q_j\}$ be a Whitney decomposition of $\mathbb{R}^N \setminus F = \Omega$. It is known that

i) $|\Omega| \leq C\|f\|_1/\alpha$,
ii) $|Q_j|^{-1} \int_{Q_j} f \leq C'\alpha, \ \forall j$.
   We define

$$g(x) := \begin{cases} f(x) & x \in F \\ |Q_j|^{-1} \int_{Q_j} f & x \in Q_j, \ j = 1, 2, \ldots \end{cases}$$

and $b := f - g$. We note that
iii) $|f(x)| \leq \alpha$ a.e. on F,
iv)

(31) $$\int_{Q_j} b = 0, \ \forall j.$$

Since $Kf = Kg + Kb$, it follows that

$$|\{x : |Kf(x)| > \alpha\}| \leq |\{x : |Kg(x)| > \frac{\alpha}{2}\}| + |\{x : |Kb(x)| > \frac{\alpha}{2}\}|$$

To estimate Kg, we note that

$$\|g\|_2^2 = \int_\Omega g^2 + \int_{D\setminus\Omega} g^2 \leq \sum_j |Q_j|^{-1} (\int_{Q_j} f)^2 + \alpha\|f\|_1 \leq$$

$$\leq C\alpha^2|\Omega| + \alpha\|f\|_1 \leq C_1\alpha\|f\|_1.$$

Applying **(A)** , we see that

(32) $$|\{Kg > \frac{\alpha}{2}\}| \leq 4\alpha^{-2}\|Kg\|_2^2 \leq C_2\alpha^{-2}\|g\|_2^2 \leq C_3\|f\|_1/\alpha.$$

It remains to estimate Kb. To do that we need the following lemma.

LEMMA 14.3.

$$|\nabla K(x,y)| \leq C|x-y|^{-N-1}, x,y \in D \text{ and } x \neq y.$$

Writing $Kb = \sum_j Kb_j$, where

$$b_j(x) := \begin{cases} b(x) & x \in Q_j, \\ 0 & x \notin Q_j, \end{cases}$$

we use (31) to deduce that

$$Kb_j(x) = \int_{Q_j} (K(x,y) - K(x,y(j)))b(y)\,dy$$

where $y(j)$ is the center of the cube Q_j.

Let $l(Q)$ be the side-length of the cube Q. Then by Lemma 14.3, and the mean value theorem

(33) $$|K(x,y) - K(x,y(j))| \leq Cl(Q_j)|x - \bar{y}(j)|^{-N-1}$$

where $\bar{y}(j)$ is a (variable) point on the line segment $[y, y(j)]$.

We shall now estimate $Kb_j(x)$ for $x \in F = \mathbb{R}^N \setminus \Omega$. By the properties of the Whitney decomposition, we see that if $x \in F$ is fixed, the set of distances $|x - y|$ as y varies in Q_j are all comparable to each other with constants of comparison independent of j. Hence

$$|Kb_j(x)| \leq Cl(Q_j) \int_{Q_j} |x-y|^{-N-1}|b(y)|\,dy \;\; \forall j \; x \in F.$$

Since for all j,

$$\int_{Q_j} |b_j(y)|\,dy \leq C\alpha|Q_j|,$$

$$l(Q_j)|Q_j| \leq C \int_{Q_j} \delta(y)\,dy$$

where $\delta(y) = \text{dist.}(y, F)$, we see that

$$|Kb_j(x)| \leq C\alpha \int_{Q_j} |x-y|^{-N-1}\delta(y)\,dy, \; x \in F,$$

and by summing

$$|Kb(x)| \leq C\alpha \int_{\mathbb{R}^N} |x-y|^{-N-1}\delta(y)\,dy, \; x \in F.$$

Applying the lemma in [33, I 2.3] we conclude that

$$\int_F |Kb(x)|\,dx \leq C\alpha|\Omega| \leq \|f\|_1.$$

It follows that

$$|\{|Kb| \geq \frac{\alpha}{2}\}| \leq |\Omega| + \frac{2}{\alpha} \int_F |Kb(x)|\,dx \leq C\|f\|_1/\alpha$$

since iii) tells us that $|\Omega| \leq C\|f\|_1/\alpha$.

Combining this result with (32), we obtain **(B)** finishing the first part of the proof of Theorem 14.1.

We will now prove **(A)** under the assumption that f is non-negative and has compact support and that $1 < p < N/(N-1)$. The result for $1 < p < \infty$ will then follow via a standard interpolation argument (cf. [33, II 2.5]).

We note that we can estimate $\int_D K(x,y)^p \, dy$ by dividing D into two parts ($|x-y| < \frac{x_1}{2}$ and $|x-y| \geq \frac{x_1}{2}$) and then using (21) and (23) to achieve the following.

$$(34) \qquad \int_D K(x,y)^p \, dy \leq C x_1^{-N(p-1)}, \quad 1 < p < N/(N-2),$$

and that if $\frac{1}{p} + \frac{1}{q} = 1$,

$$|Kg(x)| \leq C\|g\|_q x_1^{-N/q}.$$

If Q is a cube with sides parallel to the coordinate axis such that $Q^\circ \subset D$, then

$$(35) \qquad \int_Q |Kg(x)| \, dx \leq C\|g\|_q \int_Q x_1^{-N/q} \, dx_1 \, dx' \leq$$

$$\leq C\|g\|_q l(Q)^{N-1} \int_0^{l(Q)} x_1^{-N/q} \, dx_1 = C\|g\|_q |Q|^{1/p}.$$

provided that $\frac{N}{q} < 1$, or equivalently, that $p < N/(N-1)$. The constant $C = C_N$ does neither depend on Q nor on g.

Let $\tilde{Q}$ be the double of Q, i.e. a cube such that q and $\tilde{Q}$ have the same centers but $l(\tilde{Q}) = 2l(Q)$. (We consider only cubes with sides parallel to the coordinate axis.)

LEMMA 14.4. *For any cube Q, there exists a constant $C = C_N$ such that*

$$\sup_{x,x' \in Q} \int_{\mathbb{R}^N \setminus \tilde{Q}} |K(x,y) - K(x',y)| \, |g(y)| \, dy \leq C M g(z(Q)),$$

where $z(Q)$ is the center of Q.

PROOF. Let $Q_n = 2^n Q$; in particular $\tilde{Q} = Q_1$. Then applying Lemma 14.3, we deduce that

$$\int_{\mathbb{R}^N \setminus \tilde{Q}} |K(x,y) - K(x',y)| \, |g(y)| \, dy =$$

$$= \sum_{n=2}^{\infty} \int_{Q_n \setminus Q_{n-1}} |K(x,y) - K(x',y)| \, |g(y)| \, dy \leq$$

$$\leq C \sum_{n=2}^{\infty} |x - x'| \int_{Q_n \setminus Q_{n-1}} (2^{n-2} l(Q))^{-N-1} |g(y)| \, dy \leq$$

$$\leq C \sum_{n=2}^{\infty} 2^{-n} l(Q)^{-N} \int_{Q_n} |g(y)| \, dy \leq$$

$$\leq C M g(z(Q)) \sum_{n=2}^{\infty} 2^{-n} = C M g(z(Q)).$$

$\square$

Corollary of Lemma 14.4.

$$\sup_{x,x' \in Q} \int_{\mathbb{R}^N \setminus \tilde{Q}} |K(x,y) - K(x',y)| \, dy \leq C.$$

LEMMA 14.5. *Let* $\gamma = (p+1)^{-1}$ *with* $1 < p < \frac{N}{N-1}$. *If*

$\epsilon \in (0,1)$ *is given and if* $f \geq 0$, $\operatorname{supp} f \subset \tilde{Q}$, $\|f\|_1 \leq \epsilon |\tilde{Q}|$, *then*

$$|\{x \in Q : Kf(x) \geq 1\}| \leq C\epsilon^{1-\gamma}|\Omega|.$$

PROOF. By using a Calderón-Zygmund decomposition (see [33, p. 17]), we find a collection of dyadic cubes $\{Q_j\}$ in D with disjoint interiors and sides parallel to the coordinate axis such that

  i) $f(x) \leq \epsilon^\gamma$ a.e. on $D \setminus \Omega, \Omega = \cup Q_j$.
  ii) For each Q_j,

$$\epsilon^\gamma \leq |Q_j|^{-1} \int_{Q_j} f \leq C\epsilon^\gamma.$$

  iii) $|\Omega| \leq C\|f\|_1\epsilon^{-\gamma}$.

Let $\tilde{\Omega} = \cup \tilde{Q}_j$ and let $y(j)$ be the center of Q_j. Let furthermore

$$g(x) := \begin{cases} f(x) & x \in D \setminus \Omega \\ |Q_j|^{-1} \int_{Q_j} f & x \in Q_j, \ j = 1, 2, \ldots \end{cases}$$

Let $b = f - g$. We note that

$$\int_D g^q \leq C\epsilon^{\gamma(q-1)}\|f\|_1.$$

Applying (35) on the cube Q, we see that

(36) $$|\{x \in \tilde{Q} : Kg \geq 1/2\}| \leq C\|g\|_q|Q|^{1/p} \leq$$

$$\leq C\epsilon^{\gamma/p}\|f\|_1^{1/q}|Q|^{1/p} \leq C\epsilon^{\gamma/p+\frac{1}{q}}|Q| = C\epsilon^{1-\gamma}|Q|.$$

In the estimates, we used our assumption on $\|f\|_1$.

From the Corollary of Lemma 14.4, we see that

$$\int_{D \setminus \tilde{\Omega}} |Kb| \leq \sum_j \int_{x \notin \tilde{Q}_j} \int_{y \in Q_j} |K(x,y) - K(x,y(j))||b(y)|\, dy\, dx \leq$$

$$C \sum_j \int_{Q_j} |b(y)|\, dy \leq C\|f\|_1.$$

By iii), $|\tilde{\Omega}| \leq C\|f\|_1\epsilon^{-\gamma}$. We conclude that

$$|\{x \in D : |Kb(x)| \geq 1/2\}| \leq C\|f\|_1 + |\tilde{\Omega}| \leq$$

$$\leq C(\epsilon|\tilde{Q}| + \epsilon^{1-\gamma}|\tilde{Q}|) \leq C\epsilon^{1-\gamma}|Q|.$$

Combining this estimate with (36), we obtain Lemma 14.5. □

LEMMA 14.6. *There is a positive constant* a *such that if* $\lambda > 0$, $0 < \epsilon < 1$, $f \geq 0$ *and a cube* Q *contains a point* x' *satisfying*

$$Kf(x') \leq \lambda,$$

then we have the following "good λ-inequality".

(37) $$|\{x \in Q : Kf(x) \geq a\lambda, Mf(x) \leq \epsilon\lambda\}| \leq C\epsilon^{1-\gamma}|Q|$$

PROOF. Assume $x_0 \in Q$ is such that $Mf(x_0) \leq \epsilon\lambda$. Let $\tilde{Q}$ be the double of Q. Then by Lemma 14.4

(38)
$$\int_{D\setminus\tilde{Q}} K(x,y)f(y)\,dy \leq$$

$$\int_{D\setminus\tilde{Q}} |K(x,y) - K(x',y)|f(y)\,dy + Kf(x') \leq C[Mf(x_0) + \lambda] \leq C\lambda.$$

Let h be f/λ on $\tilde{Q}$ and 0 elsewhere. Then

$$\|h\|_1 = \lambda^{-1}\int_{\tilde{Q}} f \leq |\tilde{Q}|Mf(x_0)/\lambda \leq \epsilon|\tilde{Q}|$$

and Lemma 14.5 gives us the estimate

$$|\{x \in Q : Kh(x) \geq 1\}| \leq C\epsilon^{1-\gamma}|Q|.$$

From (38), we see that

$$\left|\left\{x \in Q : K\left(\frac{f}{\lambda} - h\right) \geq C + 1\right\}\right| = 0.$$

It follows that if $a > C + 2$ in (38),

$$|\{x \in Q : Kf \geq a\lambda\}| \leq |\{x \in Q : Kh \geq 1\}| +$$

$$+\left|\left\{x \in Q : K\left(\frac{f}{\lambda} - h\right) \geq a - 1\right\}\right| = |\{x \in Q : Kh \geq 1\}| \leq C\epsilon^{1-\gamma}|Q|.$$

We have proved that (37) holds. $\square$

Let us now finally prove **(A)**.

PROOF. Since
$$\|Mf\|_p \leq A(p)\|f\|_p, \quad 1 < p < \infty,$$
(cf. [33, I.1.3]) it suffices to prove that
$$\|Kf\|_p \leq C(p)\|Mf\|_p.$$

If $\|Mf\|_p$ is infinite, there is nothing to prove. We assume that $\|Mf\|_p$ is finite. Since f is non-negative and has compact support,

$$0 \leq Kf(x) \leq C|x|^{-N} \leq CMf(x), \quad x \text{ large}.$$

Hence $\|Kf\|_p$ is finite. Let $\lambda > 0$ and let $\{Q_j\}$ be a Whitney decomposition of the set $\{Kf \geq \lambda\}$.

We note that there is a constant $b > 1$ such that the cube bQ_j with the same center as Q_j and the side-length expanded b times meets the set $\{Kf < \lambda\}$ for all j.

It follows from Lemma 14.6 that

$$|\{x \in Q_j : Kf(x) \geq a\lambda, Mf(x) \leq \epsilon\lambda\}| \leq C\epsilon^{1-\gamma}|Q_j|.$$

Summing over j, we obtain

(39)
$$|\{Kf \geq a\lambda, Mf \leq \epsilon\lambda\}| \leq C\epsilon^{1-\gamma}|\{Kf \geq \lambda\}|.$$

Let $e(\lambda) = |E_\lambda|$ and $m(\lambda) = |M_\lambda|$ where

$$E_\lambda := \{Kf \geq a\lambda, Mf \leq \epsilon\lambda\},$$

$$M_\lambda := \{Mf > \epsilon\lambda\}.$$

Then by (39),

$$\|Kf\|_p^p = \int_0^\infty (a\lambda)^p \, d(-(e(\lambda) + m(\lambda))) = a^p \int_0^\infty p\lambda^{p-1} \, (e(\lambda) + m(\lambda)) \, d\lambda \leq$$

$$\leq Ca^p \epsilon^{1-\gamma} \int_0^\infty p\lambda^{p-1} |\{Kf \geq \lambda\}| \, d\lambda + (a/\epsilon)^p \|Mf\|_p^p =$$

$$Ca^p \epsilon^{1-\gamma} \|Kf\|_p^p + (a/\epsilon)^p \|Mf\|_p^p.$$

Choosing $\epsilon > 0$ so small that

$$Ca^p \epsilon^{1-\gamma} < 1/2,$$

we obtain

$$\|Kf\|_p^p \leq 2(a/\epsilon)^p \|Mf\|_p^p$$

which finishes the proof of (**A**) and hence the proof of Theorem 14.1. □

COROLLARY 14.7. *Let E be minimally thin at infinity in D. Then*

$$\int_E (1 + |x|)^{-N} \, dx < \infty.$$

History. Beurling proved his result in the planar case. In the general case ($N \geq 2$), Mazja, Dahlberg and Sjögren have showed the above result.

PROOF. As in the proof of Theorem 11.3, we see that there exists a measure μ on D such that $\mu = \sum_1^\infty \lambda_n$ where λ_n is the fundamental distribution on $E_n = E \cap I_n$, i.e.

$$G\lambda_n(x) = x_1, \quad \text{q.e. on } E_n.$$

Furthermore, $G\mu$ is defined.

Let $H_n := \{x \in D : G\lambda_n(x) \geq x_1\}$. According to Corollary 14.2,

$$|H_n| \leq C \int_E x_1 \, d\lambda_n(x) = C\gamma(E_n).$$

We note that $E \subset \cup H_n$ (except possibly for a set of capacity zero). It follows that

$$\int_E (1 + |x|)^{-N} \, dx \leq C \sum_1^\infty 2^{-nN} |H_n| \leq C \sum_1^\infty \gamma(E_n) 2^{-nN}$$

which is known to be finite (cf. Theorem 11.3). □

15. Minimal thinness, rarefiedness and ordinary capacity

The Wiener-type conditions in Theorems 11.3 and 13.1 are expressed in terms of Green energy γ or Green mass λ' of subsets of annuli. Can we replace them by ordinary capacity of subsets of Whitney cubes in a half-space D?

In the case $N = 2$, W.K. Hayman has shown that this is possible in the case of rarefied sets: he gave a talk on the subject at the BMO-seminar in Joensuu, Finland, Aug. 17–18, 1987 (cf. [22, Theorem 7.34]). In the general case, the answer is given in [15] (there is an additional remark in [16, Section 1]).

The results below hold for $N \geq 2$. For simplicity, proofs are given assuming that $N \geq 3$.

We start with the problem of repacing γ by ordinary capacity. We write D as the union of disjoint, half–open Whitny cubes $\{Q_k\}$ with sides parallel to the coordinate axis. If $Q \in \{Q_k\}$, we have

$$d(Q) = \text{dist}(Q, \partial Q) \approx \text{diameter of } Q.$$

The constants of comparison depend only on the dimension N.

To a Whitney cube Q_k, we associate numbers $\{t, r, R\} = \{t_k, r_k, R_k\}$, where t is the distance from the centre x_Q of Q to ∂D, $2r$ is the side–length of the cube and $R = |x_Q|$. We shall also need $\theta = \arccos(t/R)$. If E is a subset of D, we write $E_k = E \cap Q_k$.

$$\tilde{Q} = \{x \in \mathbb{R}^N : x - x_Q = 2(y - x_Q) \text{ for some } y \in Q\}$$

is the double of Q.

THEOREM 15.1. *The set $E \subset D$ is minimally thin at infinity in D if and only if*

$$\sum (\cos \theta_k)^2 (\log(4t_k/c(E_k)))^{-1} < \infty, \quad N = 2 \tag{40}$$

$$\sum (\cos \theta_k)^2 c(E_k) R_k^{2-N} < \infty, \quad N \geq 3. \tag{41}$$

We quote the following standard lemmas without proofs (cf. [15]).

LEMMA 15.2. *Let Q be a Whitney cube in $D \subset \mathbb{R}^2$ and let E be a subset of Q. Then we have*

$$\lambda(E) = \lambda'(E) \approx t \left(\log(4t/c(E)) \right)^{-1}, \tag{42}$$

$$\gamma(E) \approx t^2 \left(\log(4t/c(E)) \right)^{-1}. \tag{43}$$

LEMMA 15.3. *Let Q be a Whitney cube in $D \subset \mathbb{R}^N$, $N \geq 3$, and let E be a subset of Q. Then we have*

$$\lambda(E) = \lambda'(E) \approx tc(E), \tag{44}$$

$$\gamma(E) \approx t^2 c(E). \tag{45}$$

We shall also need (cf. [14, pp. 404–405])

LEMMA 15.4. *Let $G\mu$ be a Green potential in D and define*

$$I(x) := \int_{\tilde{Q}_k} G(x, y) \, d\mu(y), \quad x \in Q_k, \; k = 1, 2, \ldots$$

$$J(x) := G\mu(x) - I(x).$$

Then there exists a measure ν_0 on ∂D and a constant C such that

$$J(x) \leq CP\nu_0(x), \quad x \in D, \tag{46}$$

$$\int_{\partial D} (1 + |y|)^{-N} \, d\nu_0(y) \leq \int_D (1 + |y|)^{-N} y_1 \, d\mu(y). \tag{47}$$

Furthermore, if the support of μ is contained in $\bar{D} \cap \{1/2 \leq |x| \leq 2\}$, the support of ν_0 is contained in $\partial D \cap \{1/2 \leq |x| \leq 4\}$ and there exists a constant C_1 such that

$$\nu_0(\partial D) \leq C_1 \int_D y_1 \, d\mu(y). \tag{48}$$

PROOF. Using a trick of Sjögren, we consider a measurable mapping $\varphi : D \to \partial D$ such that

(49)
$$|x - \varphi(x)| < 2x_1, \quad x \in D,$$

(50)
$$|x| \leq \varphi(x) \leq 2|x|, \quad x \in D.$$

One possibility is to choose

$$\varphi(x) = (0, x_2, \dots, x_{p-1}, x_p + x_1 \, sgn(x_p)).$$

Let $\mathcal{F}$ be the class of continuous functions on ∂D which is such that

$$\|f\| = \sup_{\partial D} (1 + |y|)^N |f(y)| < \infty.$$

We consider the linear functional

$$L(f) = \int_D f(\varphi(y)) y_1 \, d\mu(y).$$

Since $\|L\| \leq \int_D y_1 (1 + |y|)^{-N} \, d\mu(y)$, there is a measure ν_0 on ∂D such that

(51)
$$L(f) = \int_{\partial D} f(z) \, d\nu_0(z), \quad f \in \mathcal{F}$$

and

(52)
$$\|L\| = \int_{\partial D} (1 + |z|)^{-N} \, d\nu_0(z) \leq \int_D y_1 (1 + |y|)^{-N} \, d\mu(y).$$

If $c > 0$ is given and $|x - y| > cx_1$, we have according to (49) that

$$|x - \varphi(y)| \leq |x - y| + 2y_1 \leq 3|x - y| + 2x_1 \leq (3 + 2c^{-1})|x - y|.$$

According to (21) and the definition of the Whitney cubes, we can find $c > 0$ such that

$$J(x) \leq \int_{|x-y| \geq cx_1} G(x, y) \, d\mu(y) \leq c' \int_{|x-y| \geq cx_1} x_1 y_1 |x - y|^{-N} \, d\mu(y) \leq$$

$$\leq Cx_1 \int_D y_1 |x - \varphi(y)|^{-N} \, d\mu(y) = Cx_1 \int_{\partial D} |x - z|^{-N} \, d\nu_0(z).$$

In the last step, we applied the representation formula (51) to the function $f(z) = x_1 |x - z|^{-N}$, $z \in D$, where $x \in D$ is fixed. Thus we have proved that

$$J(x) \leq CP\nu_0(x), \quad x \in D$$

and that (47) holds. (48) is a direct consequence of (47). $\square$

LEMMA 15.5. *Let E be a subset of $D \cap \{1/2 \leq |x| \leq 4\}$ and let $\{Q_k\}$ be those Whitney cubes in D which intersect this half–annulus . Let $E_k = E \cap Q_k$. Then there is an absolute constant C such that*

(53)
$$\sum \gamma(E_k) \leq C\gamma(E).$$

Remark. We know that capacity is subadditive. Inequality (53) says that Green energy γ is almost additive for decompositions of this type, or in the terminology of Aikawa "quasiadditive". In Section 16, we shall say more about quasiadditivity.

PROOF. We consider the case $N \geq 3$. with notation as in Lemma 15.4, we have

$$\hat{R}_{x_1}^E = G\mu(x) = I(x) + J(x).$$

Let ν_0 be the measure on ∂D constructed in Lemma 15.4. We claim that we can find $\alpha > 0$ such that for each k either

(54) $$J(x) \leq x_1/2 \quad \text{for all } x \in Q_k$$

or

(55) $$P\nu_0(x) > \alpha x_1 \quad \text{for all } x \in Q_k.$$

To prove this, we note that if $P\nu_0(x) \leq \alpha x_1'$ for some $x' \in Q_k$, there exists an absolute constant C_1 such that $P\nu_0(x) \leq C_1 \alpha x_1$ for all $x \in Q_k$. Choosing $\alpha = (2CC_1)^{-1}$ with C chosen as in (46), we deduce that (54) holds.

For indices k such that (54) holds, we have

(56) $$I(x) > x_1/2, \quad x \in E_k,$$

and thus that

(57) $$\hat{R}_{x_1}^E(x) \leq 2I(x), \quad x \in E_k.$$

Integrating (57) with respect to the fundamental distribution λ_k on E_k, we obtain

$$\gamma(E_k) \leq 2 \int_{\hat{Q}_k} y_1 \, d\mu(y),$$

and

$$\sum' \gamma(E_k) \leq \text{Const.} \int x_1 \, d\mu(x) = \text{Const.} \gamma(E),$$

where the prime indicates that we sum over those indices k for which (54) holds.

Let $\sum''$ denote that we sum over those indices k for which (55) holds. From (45), Corollary 14.2 and (48), we deduce that

$$\sum'' \gamma(E_k) \leq \sum'' \gamma(Q_k) \leq \text{Const.} \sum'' r_k^N \leq$$

$$\leq \text{Const.} |\{x \in D : P\nu_0(x) > \alpha x_1\}| \leq$$

$$\leq \text{Const.} \|\nu_0\| \leq \text{Const.} \int y_1 \, d\mu(y) = \text{Const.} \gamma(E).$$

Combining these two estimates, we obtain Lemma 15.5. $\square$

PROOF OF THEOREM 15.1. We give the details assuming that $N \geq 3$.
Let us first assume that (41) holds. According to Lemma 15.3, we have

$$R_k^{-N} \gamma(E_k) \approx t_k^2 c(E_k) R_k^{-N} = (\cos \theta_k)^2 c(E_k) R_k^{2-N}.$$

If $E^{(n)} = E \cap \{2^n \leq |x| \leq 2^{n+1}\}$, we use the subaddititvity of γ and see that

$$\gamma(E^{(n)}) \leq \sum \gamma(E_k),$$

where we sum over all k such that Q_k intersects $E^{(n)}$. Hence

$$\sum 2^{-nN} \gamma(E^{(n)}) \leq \text{Const.} \sum (\cos \theta_k)^2 c(E_k) R_k^{2-N} < \infty.$$

From Theorem 11.3, we see that E is minimally thin at infinity.

Conversely, assume that E is minimally thin at infinity. Let us assume that the Whitney constants have been chosen in such a way that the Whitney cubes intersecting

$E^{(n)}$ are contained in $L_n = I_{n-1} \cup I_n \cup I_{n+1}$ (cf. Section 11.2 for notation.) The Green energy of the set $E^{(n)}/2^n$ is $2^{-n}\gamma(E^{(n)})$. Applying Lemma 15.5, we deduce that

$$\sum_k \gamma(E_k)2^{-n} \leq \text{Const. } 2^{-n}\gamma(E^{(n-1)} \cup E^{(n)} \cup E^{(n+1)}\cup),$$

where we sum over all indices k such that Q_k intersects $E^{(n)}$. Summing over n and using the Wiener type criterion in Theorem 11.3 and Lemma 15.3, we obtain (41). This finishes the proof of Theorem 15.1. □

The analogue of Lemma 15.5 with γ replaced by the Green mass λ' is not true (a weak analogue is given in Lemma 15.7 below). What we can prove is the following.

THEOREM 15.6. *A set $E \subset D$ is rarefied at infinity in D if and only if E can be decomposed as $E' \cup E''$, where*

$$(58) \qquad \sum \cos \theta_k (\log(4t_k/c(E_k')))^{-1} < \infty, \quad N = 2,$$

$$(59) \qquad \sum \cos \theta_k c(E_k') R_k^{2-N} < \infty, \quad N \geq 3,$$

(here $E_k' = E' \cap Q_k$) and E'' has a covering $\cup B(X_i, r_i)$ of balls centered at $X_i \in \partial D$ with radii r_i such that

$$(60) \qquad \sum (r_i/X_i)^{N-1} < \infty.$$

(cf [15, Th. 1], [3, Th. 3.2])

PROOF. Let us assume that E is rarefied at infinity in D. According to Definition 12.4, there exists a superharmonic function $u = G\mu + P\nu$ such that $u(x) \geq |x|$ on E. Writing $G\mu = I + J$ as in Lemma 15.4, we have $u = I + (J + P\nu)$, and there exists a measure ν_1 on ∂D such that

$$(61) \qquad (J + P\nu)(x) \leq P\nu_1(x), \quad x \in D,$$

$$(62) \qquad \int_{\partial D}(1 + |y|)^{-N} d\nu_1(y) \leq$$

$$\leq C[\int_D (1 + |y|)^{-N} y_1 \, d\mu(y) + \int_{\partial D}(1 + |y|)^{-N} d\nu(y)] < \infty.$$

As usual

$$(63) \qquad E \subset \{x \in D : u(x) \geq |x|\} \subset$$

$$\subset \{x \in D : I(x) \geq |x|/2\} \cup \{x \in D : P\nu_1(x) \geq |x|/2\}$$

$$:= E' \cup E''.$$

We shall prove that E' and E'' satisfy the conditions of Theorem 15.6. We give details in the case $N \geq 3$.

We discuss first E' and start with

LEMMA 15.7. *Let μ_0 be a measure concentrated on*

$$D \cap \{1/2 \leq |x| \leq 4\} \quad \text{and define}$$

$$I_0(x) := \int_{\tilde{Q}_k} G(x,y)\,d\mu_0(y), \quad x \in Q_k.$$

If $\alpha > 0$ is given, $F = \{x \in D : I_0(x) > \alpha\}$ and $F_k = F \cap Q_k$, then

(64)
$$\sum \lambda'(F_k) \leq \text{Const.}\alpha^{-1} \int y_1\,d\mu_0(y).$$

PROOF. We have

$$\hat{R}_1^{F_k}(x) \leq \alpha^{-1} I_0(x), \quad x \in F_k.$$

Integrating with respect to the fundamental distribution $d\lambda_k$ on F_k, we obtain

$$\lambda(F_k) \leq \alpha^{-1} \int_{\tilde{Q}_k} y_1\,d\mu_0(y).$$

Since $\lambda(F_k) = \lambda'(F_k)$ (cf. Lemma 12.3), we have

$$\sum \lambda'(F_k) \leq \text{Const.}\alpha^{-1} \int y_1\,d\mu_0(y)$$

which concludes the proof of Lemma 15.4. $\square$

By assumption, E' is rarefied at infinity in D. As in the proof of Theorem 15.1, we consider the set $(E' \cap I_n)2^{-n}$. The Greeen mass of this set is $\lambda'(E \cap I_n)2^{-n(N-1)}$. Applying Lemma 15.7 with

$$d\mu_0(y) = 2^{-n(N-1)}\,d\mu(y \cdot 2^n), \quad 1 \leq |y| \leq 2,$$

we deduce that

$$\sum \lambda'(E' \cap Q_k) \leq \text{Const.}2^{-n} \int_{L_n} y_1\,d\mu_0(y)$$

where $L_n = I_{n-1} \cup I_n \cup I_{n+1}$ and we sum over all k such that Q_k intersects I_n. Hence

$$\sum \lambda'(E_k') R_k^{1-N} \leq \text{Const.} \int y_1(1+|y|)^{-N} d\mu(y) < \infty.$$

Applying Lemma 15.3, we obtain (59).

Next, we consider E'' defined by (63).

LEMMA 15.8. *Let τ be a measure supported on $\partial D \cap \{1/2 \leq |x| \leq 4\}$. Then the set*

$$H_\beta := \{x \in D : P\tau(x) > \beta\}$$

can be characterized in the following way: there exists an open set $O_\beta \in \partial D$ such that

$$H_\beta \subset D \setminus (\bigcup_{z \in \partial D \setminus O_\beta} \Gamma(z)) =: D \setminus \Omega_\beta,$$

where $\Gamma(z) = \{x \in D : 2x_1 > |x - z|\}$ is a cone in D with vertex at $z \in \partial D$ and H_β can be covered by a union of N-dimensional balls with centers on ∂D and radii $\{r_i\}$ such that

$$\sum r_i^{N-1} \leq C|O_\beta| \leq C'\|\tau\|/\beta,$$

where $|\cdot|$ denotes $(N-1)$-dimensional measure on ∂D.

PROOF. We introduce the maximal function

$$N\tau(z) := \sup_{x \in \Gamma(z)} P\tau(x), \quad z \in \partial D,$$

and the relatively open set

$$O_\beta = \{z \in \partial D : N\tau(z) > \beta\}.$$

Without loss of generality, we assume that $\beta = 1$ and write $O = O_1$ and

$$\Omega = \bigcup_{z \in \partial D \setminus O} \Gamma(z).$$

To each $x \in \Omega$, there exists $z \in \partial D \setminus O$ such that $x \in \Gamma(z)$ and

$$P\tau(x) \le N\tau(z) \le 1.$$

Hence the set $H = H_1$ is contained in $D \setminus \Omega$.

It is well known that there exists a constant C only depending on the dimension N such that

$$|O| \le C\|\tau\|.$$

(cf. [33, Th. 1 p. 197 and Th. 1 p. 5]) We can cover O by $(N-1)$-dimensional Whitney balls $\{B_i\}$ with radii $\{r_i\}$ in such a way that

$$\sum r_i^{N-1} \le C_1|O| \le CC_1\|\tau\|.$$

This terminology means that we have $r_i \approx d(B_i, \partial O)$ for all i. If $(x_1, z) \in D \setminus \Omega$ and $z \in O$ belongs to a Whitney ball B of radius r, we have

$$d(z, \partial O) \le C_0 r.$$

It follows that $x_1 \le C'r$. Hence the N-dimensional balls with centers at $\{x_{B_i}\}$ and with radii $\{2C'r_i\}$ will cover $D \setminus \Omega$. We have proved Lemma 15.8. $\square$

Let E'' be defined by (63). i.e.

$$E'' = \{x \in D : P\nu_1(x) > |x|/2\}.$$

We recall that $I_n = \{x \in D : 2^n \le |x| \le 2^{n+1}\}$ and that $L_n = I_{n-1} \cup I_n \cup I_{n+1}$. Arguing as in the proof of Theorem 13.1, we consider

$$A(x) = \int_{L_n} P(x, y) \, d\nu_1(y).$$

If $x \in I_n \cap E''$ and n is large, we have $A(x) \ge |x|/4$. Changing the scale, we define

$$d\tau(y') = 2^{-nN} d\nu_1(y' \cdot 2^n),$$

with support on $\partial D \cap \{1/2 \le |y'| \le 4\}$ which is such that

$$P\tau(x) \ge 1/4, \quad x \in (I_n \cap E'')2^{-n}.$$

Applying Lemma 15.8 and (62), we obtain (60). Thus a set E which is rarefied at infinity in D can be decomposed as in Theorem 15.6.

Conversely, assume that a set $E \subset D$ has a decomposition $E' \cup E''$ as in Theorem 15.6. Applying Theorem 13.1, we deduce that E' is rarefied at infinity in D (cf. (44) and the proof of Lemma 15.7).

To prove that E'' is rarefied at infinity in D, we shall use (60) to construct a measure ν on ∂D such that $P\nu(x)$ dominates $|x|$ on E''. Once this is done, our claim follows from Definition 12.4.

Let B be a ball with center $x_B \in \partial D$ and with radius r. It is easy to see that there exists a positive constant α such that

$$\int_{B \cap \partial D} x_1 |x - y|^{-N} dy \geq \alpha, \quad x \in B \cap D.$$

Let $\mathcal{X}_i$ be the characteristic function of $B(X_i, r) \cap \partial D$. If dy denotes surface measure on ∂D, we define

$$d\nu(y) = 2 \sum |X_i| \mathcal{X}_i(y) \, dy.$$

By assumption

$$\sum (1 + |y|)^{-N} d\nu(y) \leq \text{ Const. } \sum \left(r_i / |X_i| \right)^{N-1} < \infty,$$

and thus $P\nu$ is defined. Furthermore, we have

$$P\nu(x) \geq 2\alpha |X_i| \geq \alpha |x|, \quad x \in B_i \cap D.$$

This concludes the proof of Theorem 15.6. $\square$

Remark.   It is not difficult to go from the criterion on minimal thinness at infinity in a half-space D given in Theorem 15.1 to a criterion on minimal thinness at a boundary point of the unit ball B (cf. [17] for the case $N = 2$; the same method works for $N \geq 3$). There are here interesting connections with a minimum principle for positive harmonic functions (cf. [5] for the case $N = 2$ and [32] and [9] for $N \geq 3$; cf. also [16]) and with recent work of Bonsall and Walsh [6], Hayman and Lyons [23], Dudley Ward [11] and Gardiner [19].)

16. Quasiadditivity of capacity

It is well known that capacity is subadditive: if $E \cup E_k$, then we have

$$c(E) \leq \sum c(E_k).$$

We say that a general capacity $c(\cdot)$ is quasiadditive with respect to a decomposition $\{Q_k\}$ of a domain Ω if there exists a constant A such that

$$\sum c(E \cap Q_k) \leq Ac(E)$$

for all subsets E of Ω.

In Lemma 15.5, we proved that Green energy is quasi-additive with respect to Whitney decompositions of a half-space D: this is a key fact in the proof of Theorem 15.1 (cf. [15]).

It has been observed by Aikawa that many different capacities have this property: the word "quasi-additive" was introduced by him in 1991. In [2], he proved that Riesz capacity is quasi-additive with respect to Whitney decompositions of domains which are components of complement of closed sets without interior points in $\mathbb{R}^N$ such as m-dimensional affine subspaces with $m < N$. If $m = N - 1$, we get half-spaces. If $m = 0$, the closed set contains one point only, and we consider decompositions of $\mathbb{R}^N \setminus \{0\}$ into spherical shells. In [4], he discusses quasi-additivity of capacity and minimal thinness.

The Green mass λ' is not quasi-additive with respect to a Whitney decomposition of D: it suffices to consider the set

$$E_0 = \{x \in D : 0 < x_1 < 1, |x_k| < 1, k = 2, 3, \ldots, N\}.$$

The Whitney decomposition is too fine for λ'. To find an alternative, assume that E is a subset of E_0 and consider

$$\hat{R}_1^E = G\mu + P\nu.$$

Writing $G\mu = I + J$ as in Lemma 15.4, we have

$$\hat{R}_1^E = I + J + P\nu$$

and there exists a measure ν_1 on ∂D such that

$$J + P\nu \le P\nu_1,$$

and

(65) $$\|\nu_1\| \le C\{\|\nu\| + \int y_1 \, d\mu\} = C\lambda'(E).$$

Arguing as in the proof of Theorem 15.6, we see that $E \subset E' \cup E''$, where

$$E' = \{x \in D : I(x) \ge 1/2\},$$

$$E'' = \{x \in D : P\nu_1(x) \ge 1/2\}.$$

Let $\{Q_k\}$ be a Whitney decomposition of D. According to Lemma 15.7,

$$\sum \lambda'(E \cap Q_k) \le \text{ Const. } \int y_1 \, d\mu \le \text{ Const. } \lambda'(E).$$

To find a decomposition of D which works for λ' and E'', we argue as in the proof of Lemma 15.8 and consider the relatively open set

$$\mathcal{O} = \{z \in \partial D : N\nu_1(z) > 1/4\},$$

and the set

$$\Omega = \bigcup_{z \in \partial D \setminus \mathcal{O}} \Gamma(z).$$

We know that $E'' \subset \Omega$ and that

(66) $$|\mathcal{O}|_{N-1} \le \text{ Const. } \|\nu_1\| \le C\lambda'(E)$$

(cf. Corollary 14.2 and (65)). Let us now decompose $\mathcal{O}$ into dyadic $(N-1)$-dimensional Whitney cubes $\{T_k'\}$: the sides of the cubes $\{h_k\}$ are such that

$$h_k \approx d(T_k', \mathbb{R}^N \setminus \mathcal{O}).$$

To each cube T_k', we associate an N-dimensional box

$$T_k = \{x \in D : x = (x_1, x'), 0 \le x_1 \le \alpha h_k, x' \in T_k'\}.$$

The constant α is chosen so that

$$\Omega \subset \bigcup T_k$$

(α depends only on the Whitney constants, the opening angle of our cones Γ and the dimension).

It is easy to see that

$$\lambda'(T_k) = \lambda(T_k) \le \text{ Const. } h_k^{N-1}$$

(cf. the proof of Lemma 12.3). We conclude that

$$\sum \lambda'(E'' \cap T_k) \le \sum \lambda'(T_k) \le \text{ Const. } \sum h_k^{N-1} =$$

$$= \text{ Const. } |\mathcal{O}|_{N-1} \le C\lambda'(E)$$

(cf. (65)). Thus the decomposition $\{T_k\}$ gives quasi-additivity of Green mass for sets E'' of the type defined in Theorem 15.6.

17. On an estimate of Carleson

If a nonnegative superharmonic function is small at a point, it can not be "big" in a "large" set near the point. Results of this type are important in the study of the corona problem. The purpose of the present sections is to indicate how our techniques can be used in this context.

We assume that Q is a relatively closed square in $D = \{x_1 > 0\}$ with base on ∂D. If $l(Q)$ denotes the side length of Q, we define

$$T(Q) := \{x \in Q : \frac{1}{2}l(Q) \leq x_1 \leq l(Q)\}.$$

THEOREM 17.1. *Let D and Q be as above and let u be a nonnegative superharmonic function in D. For $b > 0$ and for $0 < \epsilon < 1$, there exists $\lambda = \lambda(b, \epsilon)$ such that for any such square*

(67)
$$\inf_{T(Q)} u(x) \leq b$$

implies

$$|E_\lambda^*|_{N-1} < \epsilon l(Q)^{N-1}$$

where E_λ^* is the orthogonal projection onto ∂D of

$$E_\lambda = \{x \in Q : u(x) \geq \lambda\}.$$

$|\cdot|_{N-1}$ *is $(N-1)$-dimensional surface measure on ∂D.*

Remark 1. It follows from the proof that there exists an absolute constant $C = C(N)$ such that

$$\lambda(b, \epsilon) \leq Cb/\epsilon.$$

Another consequence of our argument is that E_λ is contained in a union of dyadic cubes such that the sum of the surface areas of these cubes is at most $C(N)\epsilon l(Q)^{N-1}$.

Remark 2. In the case $N = 2$ and when u is the logarithm of the modulus of an analytic function, this is a result of Carleson (cf. [20, Theorem VIII.3.2, pp 334 and 367]). Carleson considers an analytic function f in the upper half-plane which is such that $||f||_\infty \leq 1$. His theorem says that if β and ϵ are given numbers in $(0, 1)$, there exists $\alpha = \alpha(\beta, \epsilon) \in (0, 1)$ such that for any square Q with basis on the real axis,

$$\sup_{T(Q)} |f(z)| \geq \beta$$

implies

$$|E_\alpha^*|_1 < \epsilon l(Q)$$

where E_α^* is the orthogonal projection onto the real axis of

$$E_\alpha(f) = \{z \in Q : |f(z)| \leq \alpha\}.$$

Putting $u(z) = \log(1/|f(z)|)$, $b = \log(1/\beta)$ and $\lambda = \log(1/\alpha)$, we obtain Theorem 17.1 in the case $N = 2$ for functions u of this type.

PROOF. A scaling argument shows that we can assume $l(Q) = 1$.
As usual (cf. Section 11.1), we have

$$u(x) = ax_1 + (G\mu + P\nu)(x),$$

where a is a nonnegative constant and μ and ν are measures concentrated on D and ∂D, respectively.

Let $\hat{x}_Q$ be the centre of the orthogonal projection of Q onto $\{x_1 = 0\}$ and define

$$\hat{Q} := \{x \in D : x - \hat{x}_Q = 2(y - \hat{x}_Q) \text{ for some } y \in Q\}.$$

We define

$$L(x) := \int_{D\setminus\hat{Q}} G(x,y) \, d\mu(y) + \int_{\partial D\setminus\partial\hat{Q}} P(x,y) \, d\nu(y)$$

and let ν_0 be the measure on ∂D constructed as in Lemma 15.4 starting from the superharmonic function

$$u(x) - L(x) = \int_{\hat{Q}} G(x,y) \, d\mu(y) + \int_{\partial\hat{Q}} P(x,y) \, d\nu(y).$$

We recall that if $\{Q_k\}$ are Whitney cubes in D contained in our given cube Q, we have

$$u(x) - L(x) = I(x) + J(x),$$

where

$$\begin{cases} I(x) = \int_{\hat{Q}_k} G(x,y) \, d\mu(y), & x \in Q_k, \quad k = 1, 2, \ldots \\ J(x) = u(x) - L(x) - I(x), & x \in Q, \end{cases}$$

where the support of ν_0 is contained say in $\partial D \cap \partial(\hat{Q})^\wedge$, and that

$$J(x) \le C_0 P\nu_0(x), \quad x \in D,$$

$$\|\nu_0\| \le C_0' \left(\int_{\hat{Q}} y_1 \, d\mu(y) + \int_{\partial\hat{Q}} d\nu(y) \right).$$

Let us assume that $\inf_{T(Q)} u(x) = b$ is attained at $w \in T(Q)$. We have

$$\{x \in Q : u(x) \ge \lambda\} \subset \bigcup_1^4 F_i,$$

where

$$F_1 = \{x \in Q : ax_1 \ge \lambda/4\},$$
$$F_2 = \{x \in Q : I(x) \ge \lambda/4\},$$
$$F_3 = \{x \in Q : J(x) \ge \lambda/4\},$$
$$F_4 = \{x \in Q : L(x) \ge \lambda/4\}.$$

To discuss F_1, we note that $a/2 \le aw_1 \le b$ and thus that $a \le 2b$. On the other hand, if $\lambda > 8b$,

$$ax_1 \le a \le 2b < \lambda/4, \quad x \in Q,$$

and F_1 will be empty.

To discuss F_4, we use a simple comparison of Poisson and Green kernels (cf. (21)), to deduce that if $L(w) \le b$ for some $w \in T(Q)$, then there exists an absolute constant $C_0 = C_0(N)$ such that

$$L(x) \le C_0 b, \quad x \in Q.$$

If $\lambda > 4C_0 b$, the set F_4 will be empty.

It remains to discuss F_2 and F_3. Again referring to (21), we note that

$$G(w,y) \geq C'_N y_1, \quad P(w,y) \geq C''_N,$$

for $w \in T(Q)$ and $y \in (\hat{Q})^-$. Since we have

$$G\mu(w) + P\nu(w) \leq b,$$

we deduce that

$$\int_{\hat{Q}} y_1 \, d\mu(y) + \int_{\partial\hat{Q}} d\nu(y) \leq C_0(N)b$$

and that

$$\|\nu_0\| \leq C_1(N)b.$$

Arguing as in Lemma 15.8, we see that if

$$E_{\lambda'}(P\nu_0) = \{x \in Q : P\nu_0(x) \geq \lambda'\},$$

then

$$|E^*_{\lambda'}(P\nu_0)|_{N-1} = \text{Const. } \|\nu_0\|/\lambda' \leq \text{Const. } b/\lambda'.$$

Combining these estimates, we deduce that

$$|E^*_{\lambda/4}(J)|_{N-1} \leq \text{Const. } b/\lambda < \epsilon/2,$$

provided that $\lambda > \text{Const. } b/\epsilon$ which gives an estimate of the contribution from F_3.

It remains to estimate the contribution from F_2. We consider $E_{\lambda/4}(I) = E'_\lambda$. Arguing as in the proof of Lemma 15.7, we see that

$$(68) \qquad \lambda'(E'_\lambda) \leq \text{Const. } \int_{\hat{Q}} y_1 \, d\mu(y)/\lambda \leq \text{Const. } b/\lambda$$

(cf. Definition 12.1).

To control E'_λ, we use a covering lemma from [14, Section 4]. Let G be a net of dyadic cubes in D with sides parallel to the coordinate axis: the length of a side of a cube in G is 2^{-n} for some $n \in \mathbf{N}$. Let $\{r_k\}$ denote side-lengths of cubes in G. We consider

$$L_{N-1}(E) = \inf \sum r_k^{N-1},$$

where the infimum is taken over all coverings of E with cubes from G (the function h in [14] is chosen as $h(s) = s^{N-1}$). In Lemma 1 in [14], it is proved that

$$(69) \qquad L_{N-1}(E) \leq \text{Const. } \lambda'(E)$$

(cf. also ([13, Lemma 5.1]).

Combining (68) and (69), we deduce that

$$|E^*_{\lambda/4}(I)| \leq L_{N-1}(E_{\lambda/4}(I)) \leq \text{Const. } b/\lambda < \epsilon/2$$

provided that $\lambda > \text{Const. } b/\epsilon$.

Chosing λ as in the discussion of the sets $F_1, \ldots, F_4$, we obtain the theorem in the case $l(Q) = 1$ and thus in general. $\square$

1.   Books on potential theory: a short list

1.1.   Classical potential theory.

L. Carleson, *Selected problems on exceptional sets.* Van Nostrand, 1967 or the reprint in Wadsworth Mathematics Series. Belmont, California, 1983.

W.K. Hayman and P.B. Kennedy, *Subharmonic functions volume 1.* Academic Press 1976.

W.K. Hayman, *Subharmonic functions volume 2.* Academic Press 1989.

L.L. Helms , *Introduction to potential theory.* Wiley–Interscience, 1969.

O. D. Kellog, *Foundations of potential theory.* Springer 1929, reprinted 1967.

N.S. Landkof, *Foundations of modern potential theory.* Springer, 1972.

J. Wermer, *Potential theory.* Springer Lecture Notes 408, 1974.

1.2.   Potential theory and function theory in the plane.

L.V. Ahlfors , *Conformal Invariants.* McGraw–Hill, 1973.

S.R. Bell, *The Cauchy transform, potential theory and conformal mapping.* CRC Press 1992.

S.D. Fisher, *Function theory on planar domains.* Wiley, 1983.

J. Garnett, *Applications of harmonic measure.* Wiley–Interscience 1986.

Chr.   Pommerenke, *Univalent functions.* Vandenhoek and Ruprecht 1975.

Chr.   Pommerenke, *Boundary behavior of conformal maps.* Springer 1992.

M. Tsuji, *Potential theory in modern function theory.* Maruzen, Tokyo 1959.

1.3.   Abstract potential theory.

J. Bliedtner and W. Hansen, *Potential theory.* Springer 1986.

M. Brelot, *On topologies and boundaries in potential theory.* Springer Lecture Notes 175, 1971. (There are many other books by Brelot.)

B. Fuglede, *Finely harmonic functions.* Springer Lecture Notes 289, 1972.

1.4.   Nonlinear potential theory.

D.R. Adams, *Lectures on L^p-potential theory.* University of Umeå, Report No. 2, 1981.

D.R. Adams and L.I. Hedberg, *Function spaces and potential theory,* Springer–Verlag 1996.

J. Heinonen, T. Kilpeläinen and O. Martio, *Non–linear potential theory.* Oxford University Press, 1992.

1.5.   Potential theory and probability.

J.L. Doob, *Classical potential theory and its probabilistic counterpart.* Springer, 1984.

S.C. Port and C.J. Stone, *Brownian motion and classical potential theory.* Academic Press, 1978.

1.6. Pluripotential theory.

E. Bedford, *Survey of pluripotential theory.* Proceedings of the special year on several complex variables. Mittag–Leffler Institute, Sweden, to appear.

U. Cegrell, *Capacities in complex analysis.* Aspects of Mathematics, Vieweg, Wiesbaden, 1988.

M. Klimek, *Pluripotential theory.* Oxford University Press, 1991.

Bibliography

[1] L.V. Ahlfors , *Conformal Invariants*. McGraw-Hill (1973).

[2] H. Aikawa, *Quasiadditivity of Riesz capacity*. Math. Scand. 69 (1991) 15-30.

[3] H. Aikawa, *Thin sets at the boundary*. Proc. London Math. Soc. (3) 65 (1992), 357-382.

[4] H. Aikawa, *Quasiadditivity of capacity and minimal thinness*. Annal. Acad. Scient. Fennicae. Ser. A.I. Mathematica 18 (1993), 65-75.

[5] A. Beurling, *A minimum principle for positive harmonic functions*. Annal. Acad. Scient. Fennicae. Ser. A.I. 372 (1965).

[6] F.F. Bonsall and D. Walsh, *Vanishing l^1-sums of the Poisson kernel, and sums with positive coefficients*. Proc. of the Edinburgh Math. Soc. 32 (1989), 431-447.

[7] L. Carleson, *Selected problems on exceptional sets*. Van Nostrand (1967).
There is also a Russian translation Л. Карлесон, Избранные проблемы теории исключительных множеств. МИР (1971).

[8] D.L. Cohn, *Measure theory*. Birkhäuser (1980).

[9] B. Dahlberg, *A minimum principle for positive harmonic functions*. Proc. of London Math. Soc. XXXIII (1976), 238-250.

[10] J.L. Doob, *Classical potential theory and its probabilistic counterpart*. Springer-Verlag (1984).

[11] N.F. Dudley Ward, *On a decomposition theorem for continuous functions of Hayman and Lyons*. New Zealand J. Math. 22 (1993), 49-59.

[12] V. Ya. Eiderman, *On the comparison of Hausdorff measure and capacity*. Algebra i Analis 3 (1991), no. 6, 174-189 (Russian); English transl. in St. Petersburg Math. J. 3 (1992), no. 6, 1367-1381.
There is also an English translation in St. Petersburg Mathematical Journal.

[13] M. Essén and H. L. Jackson *On the covering properties of certain exceptional sets in a half-space*. Hiroshima Mathematical Journal 10:2 (1980) 233-262.

[14] M. Essén , H. L. Jackson and P. J. Rippon *On minimally thin and rarefied sets in $\mathbb{R}^p$, $p \geq 2$*. Hiroshima Mathematical Journal 15 (1985) 393-410.

[15] M. Essén, *On Wiener conditions for minimally thin and rarefied sets*. "Complex Analysis", ed. J. Hersch and A. Huber, Birkhäuser Verlag, Basel 1988, pp. 41-50.

[16] M. Essén, *On minimal thinness, boundary behavior of positive harmonic functions and quasiadditivity of capacity*. Proceedings of the International Conference on Complex Analysis at the Nankai Institute of Mathematics, 1992, pp. 59-68. International Press, Cambridge MA (1994).

[17] M. Essén *On Minimal Thinness, Boundary Behavior of Positive Harmonic Functions and Quasiadditivity of Capacity* Proc. Edinburgh Math. Soc 36 (1992), 87-106.

[18] K.J. Falconer, *The geometry of fractal sets.* Cambridge Univ. Press (1985).

[19] S.J. Gardiner, *Sets of determination for harmonic functions.* Trans. Amer. Math. Soc. 338 (1993), 233-243.

[20] J.B. Garnett, *Bounded analytic functions.* Academic Press, New York (1980).

[21] V.P. Havin and V.G. Maz'ja , *Nonlinear potential theory.* Uspehi Mat. Nauk 27 (6) (1972), 67-138.

[22] W.K. Hayman, *Subharmonic functions volume 2.* Academic Press (1989).

[23] W.K. Hayman and T.J. Lyons, *Bases for positive continuous functions.* J. London Math. Soc. (2) 42 (1990), 292-308.

[24] L.I. Hedberg, *Approximation by harmonic functions, and stability of the Dirichlet problem.* Expositiones Math. 11 (1993) 193-259.

[25] J. Heinonen, T. Kilpeläinen and O. Martio, *Non-linear potential theory.* Oxford University Press, (1992).

[26] L.L. Helms , *Introduction to potential theory.* Wiley–Interscience, (1969).

[27] J.-P. Kahane and R. Salem, *Ensambles parfaits et séries trigonometriques.* Hermann, Paris (1963).

[28] O.D. Kellog, *Foundations of potential theory.* Springer (1929), reprinted 1967.

[29] N.S. Landkof, *Foundations of modern potential theory.* Springer, (1972).

[30] J. Lelong-Ferrand, *Etude au voisinage de la frontiére des fonctions surharmoniques positive dans un demi-espace.* Ann. Sci. École Norm. Sup. (3) 66 (1949), 125-159.

[31] J.L. Lewis , *Some applications of Riesz capacities.* Complex Variables 12 (1989), 237-244.

[32] V.G. Maz'ya, *Beurling's theorem on a minimum principle for positive harmonic functions.* (Russian), Zapiski Nauchnyky Seminarov LOMI 30 (1972), 76-90; (English translation), J. Soviet Math. 4 (1975), 367-379.

[33] E. Stein, *Singular Integrals and Differentiability Properties of Functions.* Princeton University Press, (1970).

[34] T. Ugaheri, *Japanese Journal of Mathematics* 20 (1950).

[35] J. Wermer, *Potential theory.* Springer Lecture Notes 408, (1974).

[36] W.P. Ziemer, *Weakly differentiable functions.* Springer–Verlag (1989).

Index

Analytic Capacity

REFERENCES GIVEN IN LECTURES BY VLADIMIR EIDERMAN

1. BOOKS

L. Carleson, *Selected problems on exceptional sets*. Van Nostrand, 1967.

T. Gamelin, *Uniform algebras*, Prentice–Hall, 1969.

J. Garnett *Analytic capacity and measure*, Lecture Notes in Math., vol. 297, Springer-Verlag, 1972.

G. M. Goluzin, *Geometric theory of functions of a complex variable*, Transl. of Math. Monogr., vol. 26, Amer. Math Soc., Providence, R. I., 1969.

W. K. Hayman and P.B. Kennedy, *Subharmonic functions volume 1*. Academic Press 1976.

N.S. Landkof, *Foundations of modern potential theory*. Springer, (1972).

M. Tsuji, *Potential theory in modern function theory*. Maruzen, Tokyo 1959.

2. PAPERS

L. Ahlfors *Bounded analytic functions*, Duke Math. J. 14 (1947), 1–11.

L. Ahlfors and A. Beurling, *Conformal invariants and function theoretic null sets*, Acta Math. 83 (1950), 101–129.

V. Ya. Eiderman, *On the comparison of Hausdorff measure and capacity*. Algebra i Analis 3 (1991), no. 6, 174–189 (Russian); English transl. in St. Petersburg Math. J. 3 (1992), no. 6, 1367–1381.

J. Garnett, *Positive length but zero analytic capacity*, Proc. Amer. Math. Soc. 21 (1970), 696–699.

V. P. Havin and S. Ya. Havinson, *Some estimates of analytic capacity*, Soviet Math. Dokl. 2 (1961), 731–734.

S. Ya. Havinson, *Analytic capacity of sets, joint nontriviality of various classes of analytic functions and the Schwarz lemma in arbitrary domains*, Mat. Sbornik 54 (1961), 3–50 (Russian); English transl. in Amer. Math. Soc. Transl., ser. 2 43, 215–266.

P. W. Jones, *Square functions, Cauchy integrals, analytic capacity, and harmonic measure*, Lecture Notes in Math., vol. 1384, Springer-Verlag, (1989), 24–68.

P. W. Jones and T. Murai, *Positive analytic capacity but zero Buffon needle probability*, Pacific J. of Math., 133 (1988), No. 1, 99–114.

P. Mattila, *A class of sets with positive length and zero analytic capacity*, Annales Acad. Sci. Fenn. Ser. A. 1. Math. 10 (1985), 387–395.

P. Mattila, *Smooth maps, null-sets for integralgeometric measure and analytic capacity*, Annals of Math. 123 (1986), 303–309.

T. Murai, *Comparison between analytic capacity and the Buffon needle probability*, Trans. Amer. Math. Soc. 304 (1987), no. 2, 501–514.

T. Murai, *Construction of H^1 functions concerning the estimate of analytic capacity*, Bull. London Math. Soc. 19 (1987), 154–160.

Ch. Pommerenke, *Über die analytische Kapazität,* Archiv der Math. 11 (1960), 270–277.

A. G. Vitushkin, *Example of a set of positive length but of zero analytic capacity,* Dokl. Akad. Nauk SSSR 127 (1959), 246–249 (Russian).

A. G. Vitushkin, *Analytic capacity of sets and problems in approximation theory,* Russian Math. Surveys 22 (1967), no. 6, 139–200.

N. X. Uy, *A removable set for Lipschitz harmonic functions,* Michigan Math. J. 37 (1990), 45–51.

3. PROBLEMS

W. K. Hayman *On Painlevé null sets,* Linear and complex analysis problem book. 199 research problems. Lecture Notes in Math., volume 1043, Springer-Verlag, Berlin etc., 1984, 491–494.

L. D. Ivanov, *On sets of analytic capacity zero,* Ibid., 498–501.

D. E. Marshall, *Removable sets for bounded analytic functions,* Ibid., 485–490.

A. G. Vitushkin, *Analytic capacity and rational approximations,* Ibid., 495–497.

Remark 1. An extensive literature has been devoted to questions connected with analytic capacity. Our list in the bibliography is far from complete. It contains mostly the literature which has been mentioned in the lectures.

Remark 2. There are also interesting reviews on Denjoy's conjecture in the sources by W. K. Hayman and D. E. Marshall given above.

Remark 3. (1995) In the following paper, it is proved that the analytic capacity of an AD–regular (Ahlfors–David) set E vanishes if an only if $H^1(E \cap \Gamma) = 0$ for all rectifiable curves Γ.

P. Mattila, M. S. Melinkov and J. Verdera *The Cauchy integral, analytic capacity, and uniform rectifiability.* To appear, Annals of Mathematics.

POTENTIAL THEORY PART II

Hiroaki Aikawa

1. Introduction

The first version of these lecture notes is based on a series of lectures given by H. Aikawa at the Department of Mathematics, University of Uppsala in the spring semester of 1993. The second version was prepared for lectures at Ochanomizu University during July 10–14 in 1995. In November 1995, a major revision was made and some sections and appendix were added.

In these lecture notes we discuss some basic materials in potential theory. After an introduction to semicontinuous functions, we start with a very general L^p-potential theory which is due to N. G. Meyers [53]. For simplicity we restrict ourselves to Euclidean space. To prove the existence of a capacity distribution, we use Clarkson's inequalities which assert that a minimizing sequence is in fact a Cauchy sequence.

The next section is devoted to the study of L^p-capacity given by a radially symmetric convolution kernel. We are particularly interested in estimates of the capacity of ball. Estimates for the Riesz and the Bessel capacity are generalized. Since our kernel is fairly general, we need the norm inequality due to Kerman-Sawyer [39]. For the convenience of the reader, we have included the proof. As an application of the estimates, we give a comparison between the L^p-capacity and the Hausdorff measure. Generally speaking, the family of sets of null capacity cannot be characterized by that of sets of null Hausdorff measure. We construct Cantor type sets which illustrate this situation. This section is based on [8].

In the succeeding section we give one more application of Kerman-Sawyer inequality. We introduce a Wolff type potential W_k^μ and observe that the L^q-norm of the convolution $k * \mu$ is estimated by the energy of W_k^μ. This observation was first established for the Bessel kernel by Hedberg and Wolff [35]. Adams and Hedberg [3, Chapter 5] used this estimate to study the behavior of capacity under a Lipschitz map and proved that the Bessel and the Riesz capacities decrease up to a multiplicative constant under a Lipschitz map. We give a generalization of their result.

One more interesting estimate is the so-called capacity strong type inequality, given first by Hansson [33], and then by Maz'ya [52, Theorem 8.2.3] and Adams [2, Theorem 1.6]. The estimate is based on the weak maximum principle of nonlinear potentials. We give an elementary proof of the capacity strong type inequality as well as the weak maximum principle. In the proof of the capacity strong type inequality, both Maz'ya and Adams used the joint measurability of the capacitary potential of the set $\{k * f > t\}$. In the present proof, we avoid the measurability and use elementary properties of series only. This is somewhat similar to the original proof of Hansson. We remark that Adams and Hedberg [3, Theorem 7.1.1] gave a beautiful short proof. The capacity strong type inequality has a lot of applications. Nagel, Rudin and Shapiro [55] used it to study the boundary behavior of Poisson integrals. Aikawa and Borichev [15] extended their result. In fact, the capacity strong type inequality is combined with the quasiadditivity of capacity and thin sets at the boundary, both of which are discussed in the following two sections.

One of the main properties of a capacity C is countable subadditivity, i.e.

$$C(E) \leq \sum C(E_j), \quad E = \bigcup E_j.$$

However, for a special decomposition we may have the reverse inequality,

$$C(E) \geq A \sum C(E_j), \quad E = \bigcup E_j,$$

where A is a constant. Such a property will be referred to as quasiadditivity. We investigate this phenomenon for the Riesz capacity and a certain Whitney decomposition. We also consider an analogue for the Green energy, which is viewed as the capacity associated with the Naïm Θ-kernel. As a result, we show refined Wiener criteria for α-thinness and minimal thinness. In retrospect, the refined Wiener criterion for minimal thinness seems to be the heart of the work of Dahlberg [24] and Sjögren [59]. See also Maz'ya [50] and [51]. The quasiadditivity of the Green energy can be extended to more wild domains such as Lipschitz and NTA domains. Hardy's inequality will play an important role. This section is based on [10] and [12].

The minimal fine limit theorem or the Fatou-Naïm-Doob theorem is one of the most beautiful theorems in potential theory (details are given in an appendix). However, we observe that this theorem is not relevant for the tangential boundary behavior of superharmonic functions. On the other hand, Nagel and Stein extended the classical Fatou nontangential limit theorem. Their approach region can contain a sequence of points with prescribed tangency. We introduce two types of exceptional set and construct some fine limit theorems based on these sets. These exceptional sets are 'thin at the boundary', but not at a particular boundary point. We note that our thin sets are natural, since a statistically minimally thin set (a set minimally thin at almost everywhere) is decomposed into the union of our two exceptional sets. Our fine limit theorems yield the Nagel-Stein theorem. This section is based on [11].

In the last main section we consider the integrability of superharmonic functions and subharmonic functions. It is an easy consequence of the Riesz decomposition theorem that a nonnegative superharmonic function is locally L^p-integrable for $0 < p < n/(n-2)$, where n is the dimension. However, its global integrability depends heavily on the regularity of the domain. This problem was first considered by Armitage for smooth domains. In fact, if $0 < p < n/(n-1)$, then every nonnegative superharmonic function on a smooth bounded domain, is globally L^p-integrable. His result was extended by Maeda and Suzuki a Lipschitz domains: their estimate of the possible values of p was not sharp. The sharp bound in dimension 2 was found by Masumoto. Lindqvist, Stegenga and Ullrich extended the results to more general domains but had to pay a price: their estimates of p was not very good. Lindqvist was able to solve the problems even for nonnegative supersolutions of a certain nonlinear equation. Here, we restrict ourselves to superharmonic functions but find the sharp value of p. The main new ingredient in our treatment is the coarea formula (cf. [13]).

For the convenience of the reader, we include in the appendices simple proofs of Choquet's capacitability theorem and of the minimal fine limit theorem (cf. Bliedtner and Hansen [20] and Brelot [21]).

Finally, the author would like to acknowledge that Professor Kaoru Hatano carefully read the manuscript and gave many helpful comments.

Acknowledgements. These notes are based on a series of lectures given at the Department of Mathematics, University of Uppsala in the spring semester of 1993. The first version of the lecture notes was completed while I was visiting the Department of Mathematics, University of Linköping. I am grateful to Professor Matts Essén, Professor Lars Inge Hedberg, Professor Vladimir Maz'ya and both the departments of Uppsala University and Linköping University. I enjoyed the valuable discussions with Dr. Alexander Borichev, Dr. Vladimir Eiderman and Dr. Torbjörn Lundh. It is my great pleasure to acknowledge the supports from the Royal Swedish Academy of Sciences and the Japan Society of Promotion of Science.

2. Semicontinuous functions

2.1. Definition and elementary properties. Let us recall the definition of continuous functions. Let u be a function on $D \subset \mathbb{R}^n$. We say u is continuous on D if for any open set U in $\mathbb{R}$, the inverse image $u^{-1}(U) = \{x \in D : u(x) \in U\}$ is (relatively) open in D. We also recall that $\{(-\infty, a), (b, +\infty) : a, b \in \mathbb{R}\}$ is an open base of $\mathbb{R}$. This means that the smallest family of sets including $\{(-\infty, a), (b, +\infty)\}$ which is closed under $\cup$ and finitely may $\cap$ is the family of all open sets of $\mathbb{R}$. Hence we observe that u is continuous if and only if $u^{-1}(-\infty, a) = \{x \in D : u(x) < a\}$ and $u^{-1}(b, +\infty) = \{x \in D : u(x) > b\}$ are open sets in D for any a and b. This leads us to the following definitions.

DEFINITION 2.1.1. Let $E \subset \mathbb{R}^n$. A function f on E is said to be *lower semicontinuous* (l.s.c.) if the following two conditions hold:
(i) $-\infty < f \le +\infty$.
(ii) For any a the set $\{x \in E : f(x) > a\}$ is a (relatively) open subset of E.

DEFINITION 2.1.2. Let $E \subset \mathbb{R}^n$. A function f on E is said to be *upper semicontinuous* (u.s.c.) if the following two conditions hold:
(i) $-\infty \le f < +\infty$.
(ii) For any a the set $\{x \in E : f(x) < a\}$ is a (relatively) open subset of E.

By $B(x, r)$ we denote the open ball with radius r and center x. We set

$$\liminf_{\substack{y \to x,\ y \ne x \\ y \in E}} f(y) = \sup_{r > 0} \left(\inf_{B(x,r) \cap E \setminus \{x\}} f \right),$$

$$\limsup_{\substack{y \to x,\ y \ne x \\ y \in E}} f(y) = \inf_{r > 0} \left(\sup_{B(x,r) \cap E \setminus \{x\}} f \right).$$

Using this notation, we obtain the following characterization of semicontinuous functions.

THEOREM 2.1.1. *Let $E \subset \mathbb{R}^n$ and let f be a function on E. Then*
(i) f is l.s.c. if and only if $\liminf_{\substack{y \to x,\ y \ne x \\ y \in E}} f(y) \ge f(x)$.
(ii) f is u.s.c. if and only if $\limsup_{\substack{y \to x,\ y \ne x \\ y \in E}} f(y) \le f(x)$.

PROOF. Let us prove (ii); (i) is left to the reader. Suppose first f is u.s.c. Let $x \in E$ be fixed and take $\alpha > f(x)$. Observe that $U = \{y \in E : f(y) < \alpha\}$ is an open set containing x. Hence we find $r > 0$ such that $B(x, r) \cap E \subset U$, which means that $\sup_{B(x,\rho) \cap E} f \le \alpha$ for $0 < \rho < r$. Hence $\limsup_{y \to x, y \in E} f(y) \le \alpha$, and by the arbitrariness of α, $\limsup_{y \to x, y \in E} f(y) \le f(x)$. Conversely, suppose $\limsup_{\substack{y \to x,\ y \ne x \\ y \in E}} f(y) \le f(x)$ for any $x \in E$. For arbitrary a we let $U = \{y \in E : f(y) < a\}$. Let us prove U is open. Suppose $x \in U$. Then $f(x) < a$ and so

$$\inf_{r > 0} \left(\sup_{B(x,r) \cap E \setminus \{x\}} f \right) \le f(x) < a.$$

Hence there exists $r > 0$ such that $f < a$ on $B(x, r) \cap E \setminus \{x\}$. The inequality obviously holds at x and so $B(x, r) \cap E \subset U$. Therefore U is open. $\square$

Exercise 2.1.1. U is open if and only if χ_U is l.s.c.; F is closed if and only if χ_F is u.s.c.

Exercise 2.1.2. f is continuous if and only if f is l.s.c. and u.s.c.

2.2. Regularizations. In this section we construct a semicontinuous function starting from an arbitrary function.

DEFINITION 2.2.1. Let f be a function on E. The function $\hat{f}(x) = \liminf_{y \to x, y \in E} f(y)$ is called the *lower regularization* of f. The function $f^*(x) = \limsup_{y \to x, y \in E} f(y)$ is called the *upper regularization* of f.

THEOREM 2.2.1. *Let f be a function on E. If the lower regularization $\hat{f}$ satisfies $\hat{f} > -\infty$, then $\hat{f}$ is l.s.c. If the upper regularization f^* satisfies $f^* < +\infty$, then f^* is u.s.c.*

PROOF. We shall prove the second assertion; the first one is left to the reader. Let a be an arbitrary number. We shall prove that $U = \{x \in E : f^*(x) < a\}$ is open. Take $x_0 \in U$. By definition we can find $r_0 > 0$ such that $\sup_{B(x_0,r_0) \cap E} f < a$. Hence, for any $x \in B(x_0, r_0) \cap E$, we have $B(x,r) \subset B(x_0, r_0)$ with $r = r_0 - |x - x_0| > 0$, and so $\sup_{B(x,r) \cap E} f < a$, whence $f^*(x) < a$. Thus $f^* < a$ on $B(x_0, r_0) \cap E$, which implies that U is open and that f^* is u.s.c. $\square$

THEOREM 2.2.2. *If $\mathcal{F}$ is a family of l.s.c. functions on E, then $F(x) = \sup_{f \in \mathcal{F}} f(x)$ is l.s.c. on E. If $\mathcal{G}$ is a family of u.s.c. functions on E, then $G(x) = \inf_{g \in \mathcal{G}} g(x)$ is u.s.c. on E.*

PROOF. We shall prove the first assertion; the second one is left to the reader. Let a be an arbitrary number and take $x_0 \in \{x \in E : F(x) > a\}$. By definition there is $f \in \mathcal{F}$ such that $f(x_0) > a$. Since f is l.s.c., there is $r > 0$ such that $f > a$ on $B(x_0, r) \cap E$. By definition $F > a$ on $B(x_0, r) \cap E$. Thus $\{x \in E : F(x) > a\}$ is open and F is l.s.c. $\square$

COROLLARY 2.2.1. *If $\{f_j\}$ is an increasing sequence of l.s.c. functions, then the limit function is l.s.c. If $\{g_j\}$ is a decreasing sequence of u.s.c. functions, then the limit function is u.s.c.*

COROLLARY 2.2.2. *Let $\{f_j\}$ be continuous functions. If the sequence $\{f_j\}$ increases to f, then f is l.s.c. If the sequence $\{f_j\}$ decreases to f, then f is u.s.c.*

2.3. Approximation. Let us prove a 'converse' of Corollary 2.2.2.

THEOREM 2.3.1. *Let f be a nonnegative l.s.c. function on E. Then there is an increasing sequence $\{f_j\}$ consisting of uniformly continuous nonnegative functions on E which converges to f.*

PROOF. We may assume that $f \not\equiv +\infty$. The construction of f_j is in one stroke; let

$$f_j(x) = \inf\{f(y) + j|x - y| : y \in E\}.$$

It is easy to see that $0 \le f_j < \infty$ and f_j is increasing. Let us prove the uniform continuity of f_j. Let $x \in E$ and $\varepsilon > 0$. By definition we find $y \in E$ such that

(2.3.1) $f_j(x) + \varepsilon > f(y) + j|x - y|$.

Take $x' \in B(x, \varepsilon) \cap E$. Since $|x' - y| < |x - y| + \varepsilon$, it follows from (2.3.1) that

$$f_j(x') \le f(y) + j|x' - y| < f(y) + j(|x - y| + \varepsilon) < f_j(x) + (j+1)\varepsilon,$$

so that $f_j(x') - f_j(x) < (j+1)\varepsilon$. Since $x' \in B(x, \varepsilon) \cap E$ if and only if $x \in B(x', \varepsilon) \cap E$ for $x, x' \in E$, we can interchange x and x' and obtain $|f_j(x') - f_j(x)| < (j+1)\varepsilon$. Thus f_j is uniformly continuous on E. Second, we prove that $f_j(x)$ increases to $f(x)$ for $x \in E$. Take $a < f(x)$. Then by l.s.c. of f we find $\delta > 0$ such that $f > a$ on $B(x, \delta) \cap E$. Hence $\inf\{f(y) + j|x - y| : y \in B(x, \delta) \cap E\} \geq a$. On the other hand if $y \in E \setminus B(x, \delta)$, then $f(y) + j|x - y| \geq j\delta \to +\infty$ as $j \to \infty$. Hence if $j\delta > a$, then $f_j(x) \geq a$. This means that $f_j(x) \uparrow f(x)$. $\square$

COROLLARY 2.3.1. *Let U be an open set and let f be a nonnegative l.s.c. function on U. Then there is an increasing sequence $\{f_j\}$ of continuous nonnegative functions on U with compact support which converges to f.*

PROOF. Let $\{\varphi_j\}$ be an increasing sequence of nonnegative functions on U with compact support converging to χ_U. Let $\{f_j\}$ be as in Theorem 2.3.1. Then $\{f_j\varphi_j\}$ is the required sequence. $\square$

Exercise 2.3.1. State the counterpart of Theorem 2.3.1 in the context of u.s.c. functions and prove it.

2.4. Vague convergence. In this subsection we consider some convergence property of the integral

$$k(x, \mu) = \int_{\mathbb{R}^n} k(x, y) d\mu(y),$$

where $k(x, y)$ is a nonnegative l.s.c. function on $\mathbb{R}^n \times \mathbb{R}^n$. It is easy to see from Fatou's lemma that $k(x, \mu)$ is l.s.c. for a nonnegative measure μ. We say that a sequence of signed measures μ_j converges vaguely to μ if

$$\int_{\mathbb{R}^n} f(x) d\mu_j(x) \to \int_{\mathbb{R}^n} f(x) d\mu(x) \quad \text{for all } f \in C_0(\mathbb{R}^n),$$

where $C_0(\mathbb{R}^n)$ denotes the family of all continuous functions with compact support on $\mathbb{R}^n$. The vague convergence is very weak.

Exercise 2.4.1. Let x_j and y_j be sequences converging to 0. Let $\mu_j = \delta_{x_j} - \delta_{y_j}$, where δ_x is the point measure at x. Prove that μ_j converges vaguely to 0.

Exercise 2.4.2. Let μ_j be nonnegative measures converging vaguely to μ. Prove that $\mu_j(K)$ are bounded for every compact subset K of $\mathbb{R}^n$.

THEOREM 2.4.1. *Let $k(x, y)$ be a nonnegative l.s.c. function on $\mathbb{R}^n \times \mathbb{R}^n$. Let μ_j be nonnegative measures converging vaguely to μ and let $x_j \to x$. Then*

$$\liminf_{j \to \infty} k(x_j, \mu_j) \geq k(x, \mu).$$

PROOF. Let $\alpha < k(x, \mu)$. Since k is l.s.c., it follows from Corollary 2.3.1 that there is an increasing sequence $k_m \in C_0(\mathbb{R}^n \times \mathbb{R}^n)$ converging to k. By the monotone convergence theorem, we find m such that $k_m(x, \mu) > \alpha$. We let $\varepsilon = k_m(x, \mu) - \alpha > 0$. Since $k_m \in C_0(\mathbb{R}^n \times \mathbb{R}^n)$, we can take a compact subset K of $\mathbb{R}^n$ such that

$$k_m(\cdot, \nu) = \int_K k_m(\cdot, y) d\nu(y) \quad \text{for any measure } \nu.$$

As observed in Exercise 2.4.2, the vague convergence of μ_j implies the boundedness of $\mu_j(K)$. Let $\mu_j(K) \leq M$. By the uniform continuity of k_m we find j_1 such that if $j \geq j_1$, then

$$|k_m(x_j, y) - k_m(x, y)| < \frac{\varepsilon}{2M} \quad \text{for all } y \in \mathbb{R}^n.$$

Also, by the vague convergence of μ_j, we find j_2 such that if $j \geq j_2$, then

$$|k_m(x, \mu_j) - k_m(x, \mu)| < \frac{\varepsilon}{2},$$

since $k_m(x, \cdot) \in C_0(\mathbb{R}^n)$. Now let $j \geq \max\{j_1, j_2\}$. Then

$$k_m(x_j, \mu_j) = k_m(x, \mu_j) + \int_K (k_m(x_j, y) - k_m(x, y)) d\mu_j(y)$$

$$\geq k_m(x, \mu) - |k_m(x, \mu) - k_m(x, \mu_j)| - \frac{\varepsilon}{2M} \mu_j(K)$$

$$> k_m(x, \mu) - \frac{\varepsilon}{2} - \frac{\varepsilon}{2} = \alpha.$$

This implies the required inequality since $\alpha < k(x, \mu)$ is arbitrary. $\square$

COROLLARY 2.4.1. *Let $k(x, y)$ be a nonnegative l.s.c. function on $\mathbb{R}^n \times \mathbb{R}^n$. Let μ_j be nonnegative measures converging vaguely to μ. Then*

$$\liminf_{j \to \infty} k(x, \mu_j) \geq k(x, \mu).$$

3. L^p capacity theory

This part is taken from N. G. Meyers [53].

3.1. Preliminaries. The theory of Meyers is very general. For simplicity, we shall restrict ourselves to the Euclidean space $\mathbb{R}^n$ of dimension n. Throughout this part we let $1 < p < \infty$ and let p' be its conjugate, i.e.

$$\frac{1}{p} + \frac{1}{p'} = 1.$$

The L^p-norm of f is given by

$$\|f\|_p = \left(\int |f|^p dx \right)^{1/p}.$$

By $C_0(\mathbb{R}^n)$ we denote the family of all continuous functions with compact support in $\mathbb{R}^n$.

LEMMA 3.1.1.

$$\|f\|_p = \sup_{\|g\|_{p'} \leq 1} \left| \int fg \, dx \right| = \sup_{\|g\|_{p'} \leq 1, g \in C_0(\mathbb{R}^n)} \left| \int fg \, dx \right|.$$

Moreover, if $f \geq 0$, then

$$\|f\|_p = \sup_{\|g\|_{p'} \leq 1, g \geq 0} \left| \int fg \, dx \right| = \sup_{\|g\|_{p'} \leq 1, g \in C_0(\mathbb{R}^n), g \geq 0} \left| \int fg \, dx \right|.$$

PROOF. Let us prove the first equality. The second follows from the denseness of $C_0(\mathbb{R}^n)$ in $L^{p'}(\mathbb{R}^n)$. The Hölder inequality says that

$$\left| \int fg \, dx \right| \leq \|f\|_p \|g\|_{p'} \leq \|f\|_p,$$

if $\|g\|_{p'} \leq 1$. Thus the middle supremum is less than or equal to $\|f\|_p$. For the converse we let

$$g = \begin{cases} \dfrac{\bar{f}}{|f|} |f|^{p-1} / \|f\|_p^{p-1} & \text{if } f(x) \neq 0, \\ 0 & \text{if } f(x) = 0, \end{cases}$$

where $\overline{f}$ is the complex conjugate of f. Observe that

$$\|g\|_{p'} = \frac{1}{\|f\|_p^{p-1}}\left(\int |f|^{(p-1)p'}dx\right)^{1/p'} = 1,$$

and

$$\int fg\,dx = \frac{1}{\|f\|_p^{p-1}}\int |f|^p dx = \|f\|_p.$$

Thus the middle supremum is not less than $\|f\|_p$. The second assertion is clear from the above argument. The lemma is proved. $\square$

We say that f_j converges to f weakly in $L^p(\mathbb{R}^n)$, denoted by $f_j \rightharpoonup f$ weakly in $L^p(\mathbb{R}^n)$, if

$$\int f_j g\,dx \to \int fg\,dx \text{ for all } g \in L^{p'}(\mathbb{R}^n).$$

LEMMA 3.1.2. If $f_j \rightharpoonup f$ weakly in $L^p(\mathbb{R}^n)$, then

$$\|f\|_p \le \liminf_{j\to\infty} \|f_j\|_p.$$

PROOF. Take an arbitrary function $g \in L^{p'}(\mathbb{R}^n)$ such that $\|g\|_{p'} \le 1$. Then the weak convergence says

$$\left|\int fg\,dx\right| = \lim_{j\to\infty}\left|\int f_j g\,dx\right| \le \liminf_{j\to\infty}\|f_j\|_p\|g\|_{p'} \le \liminf_{j\to\infty}\|f_j\|_p.$$

Since g was arbitrary, Lemma 3.1.1 yields the required inequality. $\square$

3.2. Definition and elementary properties. Let $k(x, y)$ be a nonnegative lower semi-continuous (l.s.c.) function on $\mathbb{R}^n \times \mathbb{R}^n$. Such a function will be called a kernel. We note that k may be infinite on the diagonal set $\{(x, x) : x \in \mathbb{R}^n\}$. For a measure ν we write

$$k(x, \nu) = \int k(x, y)d\nu(y).$$

If ν has the density f with respect to Lebesgue measure, then we write $k(x, f)$. We define $k(\nu, y)$ and $k(f, y)$, similarly. We write

$$k(\mu, \nu) = \int k(x, y)d\mu(x)d\nu(y).$$

DEFINITION. For any set A we define

$$C_{k,p}(A) = \inf\{\|f\|_p^p : f \ge 0, k(\cdot, f) \ge 1 \text{ on } A\}.$$

The above infimum means ∞ if there is no feasible f.

We give some elementary properties of $C_{k,p}$.

THEOREM 3.2.1. (i) $C_{k,p}(\emptyset) = 0$.
(ii) $C_{k,p}$ is monotone; if $A \subset B$, then $C_{k,p}(A) \le C_{k,p}(B)$.
(iii) $C_{k,p}$ is countably subadditive (c.s.a.);

$$C_{k,p}\left(\bigcup_{j=1}^{\infty} A_j\right) \le \sum_{j=1}^{\infty} C_{k,p}(A_j).$$

(iv) $C_{k,p}$ is an outer capacity;

$$C_{k,p}(A) = \inf_{\substack{U \supset A \\ U \text{ is open}}} C_{k,p}(U).$$

PROOF. By definition (i) and (ii) readily follow. Let us prove (iii). We may assume that $\sum_{j=1}^{\infty} C_{k,p}(A_j) < \infty$. In particular, $C_{k,p}(A_j) < \infty$ for each j. For an arbitrary $\varepsilon > 0$ we can find f_j such that $k(\cdot, f_j) \geq 1$ on A_j and

$$\|f_j\|_p^p \leq C_{k,p}(A_j) + 2^{-j}\varepsilon.$$

Let $f(x) = \sup_j f_j(x)$. Then

$$f(x)^p \leq \sum_{j=1}^{\infty} f_j(x)^p \text{ for all } x,$$

$$k(\cdot, f) \geq k(\cdot, f_j) \text{ for all } j$$

Hence $k(\cdot, f) \geq 1$ on $\bigcup_{j=1}^{\infty} A_j$ and the definition yields

$$C_{k,p}\left(\bigcup_{j=1}^{\infty} A_j\right) \leq \|f\|_p^p \leq \int \sum_{j=1}^{\infty} f_j(x)^p = \sum_{j=1}^{\infty} \|f_j\|_p^p \leq \sum_{j=1}^{\infty} C_{k,p}(A_j) + \varepsilon.$$

Thus the required inequality follows.

For the proof of (iv) we may assume that $C_{k,p}(A) < \infty$. Let $0 < \varepsilon < 1$. By definition we find $f \geq 0$ such that $k(\cdot, f) \geq 1$ on A and

$$\|f\|_p^p \leq C_{k,p}(A) + \varepsilon.$$

Let $f_\varepsilon = (1 - \varepsilon)^{-1} f$. Then $k(\cdot, f_\varepsilon) \geq (1 - \varepsilon)^{-1} > 1$ on A. From the l.s.c. of k we observe that $U = \{x : k(x, f_\varepsilon) > 1\}$ is an open set including A. By definition

$$C_{k,p}(U) \leq \|f_\varepsilon\|_p^p = (1 - \varepsilon)^{-p}\|f\|_p^p \leq (1 - \varepsilon)^{-p}(C_{k,p}(A) + \varepsilon) \to C_{k,p}(A) \text{ as } \varepsilon \to 0.$$

Thus (iv) follows. $\square$

3.3. Convergence properties.

LEMMA 3.3.1. *(Strong convergence) Let $f_j \to f$ in $L^p(\mathbb{R}^n)$. Then there is a subsequence f_j' such that*

$$k(\cdot, f_j') \to k(\cdot, f) \quad C_{k,p}\text{-}a.e.,$$

where $C_{k,p}$-a.e. means that the property holds outside a set of $C_{k,p}$-capacity null.

PROOF. Fix $\varepsilon > 0$ and let

$$A_{\varepsilon,j} = \{x : |k(x, f_j) - k(x, f)| \geq \varepsilon\}.$$

Then by definition

$$C_{k,p}(A_{\varepsilon,j}) \leq \varepsilon^{-p}\|f_j - f\|_p^p \to 0 \text{ as } j \to \infty.$$

Let $\varepsilon = 2^{-i}$ and choose $j(i)$ such that

$$C_{k,p}(A_{\varepsilon,j(i)}) < 2^{-i}.$$

By $A(i)$ we denote $A_{\varepsilon,j(i)}$. We show that the sequence $f_{j(i)}$ has the required property. In fact the upper limit set

$$A = \bigcap_{l=1}^{\infty} \bigcup_{i=l}^{\infty} A(i)$$

has null $C_{k,p}$-capacity, since the monotonicity and the c.s.a. imply

$$C_{k,p}(A) \leq \sum_{i=l}^{\infty} C_{k,p}(A(i)) \leq \sum_{i=l}^{\infty} 2^{-i}$$

for all l. If $x \notin A$, then there exists $l = l(x)$ such that $x \notin A(i)$ for $i \geq l$, i.e.

$$|k(x, f_{j(i)}) - k(x, f)| < 2^{-i}.$$

This implies $k(x, f_{j(i)}) \to k(x, f)$ as $i \to \infty$. Thus the lemma follows. $\square$

Obviously, weak convergence implies vague convergence. Hence we have the following lemma.

LEMMA 3.3.2. *(Weak convergence 1) Let $f_j \to f$ weakly in $L_+^p(\mathbb{R}^n)$. Then*

$$\liminf_{j \to \infty} k(\cdot, f_j) \geq k(\cdot, f) \text{ everywhere on } \mathbb{R}^n.$$

Moreover, if $x_j \to x$, then

$$\liminf_{j \to \infty} k(x_j, f_j) \geq k(x, f).$$

LEMMA 3.3.3. *(Weak convergence 2) Let $f_j \to f$ weakly in $L_+^p(\mathbb{R}^n)$. Then*

$$\liminf_{j \to \infty} k(\cdot, f_j) = k(\cdot, f) \quad C_{k,p}\text{-a.e. on } \mathbb{R}^n.$$

PROOF. By Lemma 3.3.2 we have only to show that

$$\liminf_{j \to \infty} k(\cdot, f_j) \leq k(\cdot, f) \quad C_{k,p}\text{-a.e. on } \mathbb{R}^n.$$

In view of the Banach-Saks theorem (see e.g. [68, III A, 27]) or the Mazur theorem (see e.g. [69, Theorem 2 in Section V.1]), we find a subsequence f_j' such that the average

$$g_j = \frac{1}{j} \sum_{i=1}^{j} f_i'$$

converges strongly to f in $L^p(\mathbb{R}^n)$. By Lemma 3.3.1 we can find a subsequence g_j' such that

$$k(\cdot, g_j') \to k(\cdot, f) \quad C_{k,p}\text{-a.e.}$$

Therefore

$$\liminf_{j \to \infty} k(\cdot, f_j) \leq \liminf_{j \to \infty} k(\cdot, f_j') \leq \liminf_{j \to \infty} k(\cdot, g_j) \leq \liminf_{j \to \infty} k(\cdot, g_j') = k(\cdot, f) \quad C_{k,p}\text{-a.e.}$$

Thus the required inequality follows and the lemma is proved. $\square$

THEOREM 3.3.1.   (i) *If $A_j \uparrow A$, then $C_{k,p}(A_j) \uparrow C_{k,p}(A)$.*
(ii) *Let K_j be compact sets. If $K_j \downarrow K$, then $C_{k,p}(K_j) \downarrow C_{k,p}(K)$.*

PROOF. For the proof of (i) we may assume that

$$\alpha = \lim_{j \to \infty} C_{k,p}(A_j) < \infty,$$

since $\alpha \leq C_{k,p}(A)$ by the monotonicity. By definition we can find $f_j \geq 0$ such that

$$k(\cdot, f_j) \geq 1 \text{ on } A_j$$
$$\|f_j\|_p^p \leq C_{k,p}(A_j) + 1/j \leq \alpha + 1.$$

Since $\{f_j\}$ is a bounded sequence in $L_+^p(\mathbb{R}^n)$, we may assume, if necessary taking a subsequence, that $f_j \to f$ weakly in $L_+^p(\mathbb{R}^n)$ (cf. [69, Theorem 1 on p.126]). By Lemma 3.1.2 we have

(3.3.1) $$\|f\|_p^p \leq \liminf_{j \to \infty} \|f_j\|_p^p = \alpha.$$

Lemma 3.3.3 yields that

$$\liminf_{j\to\infty} k(\cdot, f_j) = k(\cdot, f) \ C_{k,p}\text{-a.e.}$$

This means that there is a subset B of A such that $C_{k,p}(A \setminus B) = 0$ and $k(\cdot, f) \geq 1$ on B. By (3.3.1) we see that $C_{k,p}(B) \leq \|f\|_p^p \leq \alpha$. Thus the subadditivity yields

$$C_{k,p}(A) \leq C_{k,p}(A \setminus B) + C_{k,p}(B) = \alpha,$$

which implies $C_{k,p}(A_j) \uparrow C_{k,p}(A)$.

It is known that (ii) holds for general outer capacities. For the completeness, we give a proof. By the monotonicity we have

$$\lim_{j\to\infty} C_{k,p}(K_j) \geq C_{k,p}(K).$$

For the reverse inequality we take an arbitrary $\varepsilon > 0$. Since $C_{k,p}$ is an outer capacity, we can find an open set $U \supset K$ such that

$$C_{k,p}(U) \leq C_{k,p}(K) + \varepsilon.$$

We see that $K_j \subset U$ for large j (See Remark below), so that $C_{k,p}(K_j) \leq C_{k,p}(K) + \varepsilon$ by the monotonicity. This means

$$\lim_{j\to\infty} C_{k,p}(K_j) \leq C_{k,p}(K) + \varepsilon,$$

and the arbitrariness of ε yields the required inequality. $\square$

Remark. Observe that $\{K_j \setminus U\}$ is a decreasing sequence of compact sets. If the statement does not hold, then these compact sets have the finite-intersection property, which yields $K \setminus U = \bigcap_{j=1}^{\infty}(K_j \setminus U) \neq \emptyset$, a contradiction.

By Theorem 3.2.1 and Theorem 3.3.1 we obtain the following fundamental theorem. This is the famous capacitability theorem of Choquet. For the convenience of reader, we present a proof in an appendix.

THEOREM 3.3.2. *All analytic sets A are capacitable, i.e.*

$$C_{k,p}(A) = \inf_{\substack{U \supset A \\ U \text{ is open}}} C_{k,p}(U) = \sup_{\substack{K \subset A \\ K \text{ is compact}}} C_{k,p}(K).$$

3.4. Capacitary distributions. Let us observe that

$$C_{k,p}(A) = \inf\{\|f\|_p^p : f \geq 0, k(\cdot, f) \geq 1 \ C_{k,p}\text{-a.e. on } A\}.$$

We shall show that if $C_{k,p}(A) < \infty$, then there exists a unique nonnegative function f for which the above infimum is attained.

DEFINITION. Let $C_{k,p}(A) < \infty$. We say that $f \geq 0$ is the $C_{k,p}$-capacitary distribution for A, if

$$\|f\|_p^p = C_{k,p}(A),$$
$$k(\cdot, f) \geq 1 \ C_{k,p}\text{-a.e. on } A.$$

We say that $k(\cdot, f)$ is the $C_{k,p}$-capacitary potential for A.

THEOREM 3.4.1. *Let $C_{k,p}(A) < \infty$. Then there is a unique $C_{k,p}$-capacitary distribution for A.*

In order to prove the theorem, we shall invoke the following inequalities, which are known as *Clarkson's inequalities*. For a proof see e.g. [5, Theorem 2.28]. Several elementary proofs are known. (The reader is encouraged to try to find such proofs in the literature.)

LEMMA 3.4.1. *(Clarkson's inequality)*
(i) Let $p \geq 2$. Then

$$\left\| \frac{f+g}{2} \right\|_p^p + \left\| \frac{f-g}{2} \right\|_p^p \leq \frac{1}{2} \|f\|_p^p + \frac{1}{2} \|g\|_p^p.$$

(ii) Let $1 < p \leq 2$. Then

$$\left\| \frac{f+g}{2} \right\|_p^{p'} + \left\| \frac{f-g}{2} \right\|_p^{p'} \leq \left(\frac{1}{2} \|f\|_p^p + \frac{1}{2} \|g\|_p^p \right)^{p'-1}.$$

PROOF OF THEOREM 3.4.1. Let $\alpha = C_{k,p}(A) < \infty$. Then there is $f_j \in L_+^p(\mathbb{R}^n)$ such that $k(\cdot, f_j) \geq 1$ on A and $\|f_j\|_p^p \to \alpha$. We show that $\{f_j\}$ is a Cauchy sequence in $L_+^p(\mathbb{R}^n)$. Note that $k(\cdot, \frac{1}{2}(f_i + f_j)) \geq 1$ on A and hence

$$\left\| \frac{f_i + f_j}{2} \right\|_p^p \geq \alpha$$

by definition. Suppose $p \geq 2$. Then Clarkson's inequality implies

$$\left\| \frac{f_i - f_j}{2} \right\|_p^p \leq \frac{1}{2} \|f_i\|_p^p + \frac{1}{2} \|f_j\|_p^p - \left\| \frac{f_i + f_j}{2} \right\|_p^p.$$

The first two terms in the right hand side tend to $\frac{1}{2}\alpha$; the last term is not less than α as observed above. Hence

$$\limsup_{j \to \infty} \left\| \frac{f_i - f_j}{2} \right\|_p^p \leq 0.$$

Thus $\{f_j\}$ is a Cauchy sequence and we can find $f \in L_+^p(\mathbb{R}^n)$ such that

$$f_j \to f \text{ strongly,}$$
$$\|f\|_p^p = \alpha = C_{k,p}(A).$$

By Lemma 3.3.1 we see that

$$k(\cdot, f) \geq 1 \ C_{k,p}\text{-a.e. on } A.$$

Thus this f is the capacitary distribution. It is easy to see that the above argument also implies the uniqueness of f. The result in the case $1 < p \leq 2$ can be proved in a similar fashion. $\square$

COROLLARY 3.4.1. *Let $A \subset B$ and $C_{k,p}(A) = C_{k,p}(B) < \infty$. Then A and B have the same capacitary distribution.*

PROOF. Let f be the capacitary distribution for B. Then $\|f\|_p^p = C_{k,p}(B) = C_{k,p}(A)$ and $k(\cdot, f) \geq 1$ on $C_{k,p}$-a.e. on B and hence on A. By the uniqueness of the capacitary distribution we see that f coincides with the capacitary distribution for A. $\square$

3.5. Dual capacity. We say that a set A is *universally measurable* if A is measurable for all nonnegative Radon measures. Here, a measure μ is called a Radon measure if μ is defined for all Borel sets and $\mu(K) < \infty$ for any compact set K. Hereafter, every measure is a Radon measure. Obviously, every Borel set is universally measurable. Moreover, by the capacitability theorem every analytic set is universally measurable. In this section we shall define another capacity $c_{k,p}$ for universally measurable sets. In general, we say that a measure μ is *concentrated on* A if $\mu(A^c) = 0$. If $\mathrm{supp}(\mu) \subset A$, then μ is concentrated on A. But the converse is not necessarily true, unless A is closed.

DEFINITION. Let A be a universally measurable set. We define

$$c_{k,p}(A) = \sup\{\|\nu\|_1 : \ \nu \text{ is concentrated on } A, \ \|k(\nu,\cdot)\|_{p'} \le 1\}.$$

A measure ν satisfying the condition in the supremum will be called a *test measure* for $c_{k,p}(A)$, or more simply for A.

THEOREM 3.5.1. $c_{k,p}$ *is an inner capacity defined over universally measurable sets. In fact, it satisfies the following:*

(i) $c_{k,p}(\emptyset) = 0$.
(ii) $c_{k,p}$ *is monotone; if* $A \subset B$, *then* $c_{k,p}(A) \le c_{k,p}(B)$.
(iii) $c_{k,p}$ *is countably subadditive (c.s.a.);*

$$c_{k,p}\left(\bigcup_{j=1}^{\infty} A_j\right) \le \sum_{j=1}^{\infty} c_{k,p}(A_j).$$

(iv)

$$c_{k,p}(A) = \sup_{\substack{K \subset A \\ K \text{ is compact}}} c_{k,p}(K).$$

Here, the sets A, B and A_j are universally measurable sets.

PROOF. By definition (i) and (ii) readily follow. Let us prove (iii). We may assume that the sets $\{A_j\}$ are disjoint. Take an arbitrary test measure for $A = \bigcup_{j=1}^{\infty} A_j$ and write ν_j for the restriction $\nu|_{A_j}$. Since our sets are disjoint, it follows that

$$\nu = \sum_{j=1}^{\infty} \nu_j.$$

Obviously, ν_j is a test measure for A_j. Hence

$$\|\nu\|_1 = \sum_{j=1}^{\infty} \|\nu_j\|_1 \le \sum_{j=1}^{\infty} c_{k,p}(A_j).$$

Since ν is an arbitrary test measure for A, we have

$$c_{k,p}(A) \le \sum_{j=1}^{\infty} c_{k,p}(A_j).$$

For (iv) we take an arbitrary test measure ν for a universally measurable set A. If K is a compact subset of A, then the restriction $\nu|_K$ is a test measure for K. Note that

$$\|\nu\|_1 = \sup_{\substack{K \subset A \\ K \text{ is compact}}} \nu(K) = \sup_{\substack{K \subset A \\ K \text{ is compact}}} \|\nu|_K\|_1,$$

whence by definition

$$\|\nu\|_1 \leq \sup_{\substack{K \subset A \\ K \text{ is compact}}} c_{k,p}(K).$$

Since ν is an arbitrary test measure for A, it follows that

$$c_{k,p}(A) \leq \sup_{\substack{K \subset A \\ K \text{ is compact}}} c_{k,p}(K).$$

The opposite inequality follows from the monotonicity. Thus we have the required equality. The theorem is proved. $\square$

DEFINITION. We define an outer capacity $c_{k,p}^*$ from $c_{k,p}$ by

$$c_{k,p}^*(A) = \inf_{\substack{U \supset A \\ U \text{ is open}}} c_{k,p}(U)$$

The outer capacity $c_{k,p}^*(A)$ can be defined for any set.

DEFINITION. Let A be a universally measurable set with $c_{k,p}(A) < \infty$. We say that ν is a $c_{k,p}$-capacitary distribution for A, if ν is a test measure for A and $\|\nu\|_1 = c_{k,p}(A)$. We say that $k(\nu, \cdot)$ is a $c_{k,p}$-capacitary potential for A.

Remark. If ν is a $c_{k,p}$-capacitary distribution for A, then $\|k(\nu, \cdot)\|_{p'} = 1$.

THEOREM 3.5.2. Every compact set K of finite $c_{k,p}$-capacity has a $c_{k,p}$-capacitary distribution for K.

PROOF. Let ν_j be test measures for K such that $\|\nu_j\|_1 \uparrow c_{k,p}(K) < \infty$. Since $\{\nu : \text{supp}(\nu) \subset K, \|\nu\|_1 \leq c_{k,p}(K)\}$ is sequentially weakly compact, we may assume, if necessary by taking a subsequence, that ν_j converges vaguely to a measure ν supported on K, i.e.

$$\int f d\nu_j \to \int f d\nu \quad \text{for all } f \in C_0(\mathbb{R}^n).$$

In particular, $\|\nu_j\|_1 \to \|\nu\|_1$, so that $\|\nu\|_1 = c_{k,p}(K)$. From the l.s.c. of k we have

$$\|k(\nu, \cdot)\|_{p'} \leq \liminf_{j \to \infty} \|k(\nu_j, \cdot)\|_{p'}.$$

The right hand side is not greater than 1 by the choice of ν_j. Thus ν is a test measure for K and hence ν is a $c_{k,p}$-capacitary distribution for K. $\square$

Remark. Let A be a closed set. If $c_{k,p}(A) < \infty$ and $c_{k,p}(\{x \in A : |x| > r\}) \downarrow 0$ as $r \uparrow \infty$, then A has a $c_{k,p}$-capacitary distribution. This is a generalization of the above theorem. For a proof we refer to [53, Theorem 13].

3.6. Duality. The capacities $C_{k,p}$, $c_{k,p}$ and $c_{k,p}^*$ have the following relationship.

THEOREM 3.6.1.

$$c_{k,p}^*(A)^p = C_{k,p}(A) \quad \text{for all sets } A;$$
$$c_{k,p}(A)^p = C_{k,p}(A) \quad \text{for all analytic sets } A.$$

For the proof of the theorem we shall use the following mini-max theorem. The mini-max theorem was first established by von Neumann in the theory of games. There are several mathematical extensions of von Neumann's mini-max theorem. B. Fuglede [30] first applied the mini-max theorem to potential theory. The following is a general mini-max theorem. For a proof we refer to Aubin [19].

LEMMA 3.6.1. *(Mini-max theorem) Let V_1 be a real linear space and let V_2 be a Hausdorff real linear space. Suppose $H_1 \neq \emptyset$ is a convex subset of V_1 and $H_2 \neq \emptyset$ is a convex compact subset of V_2. Assume that $\Phi(x_1, x_2)$ is a real valued function on $H_1 \times H_2$ such that*

$$\Phi(x_1, x_2) > -\infty,$$

$$\Phi(\cdot, x_2) \text{ is concave for all fixed } x_2 \in H_2,$$

$$\Phi(x_1, \cdot) \text{ is convex and l.s.c. for all fixed } x_1 \in H_1.$$

Then

$$\sup_{x_1 \in H_1} \inf_{x_2 \in H_2} \Phi(x_1, x_2) = \inf_{x_2 \in H_2} \sup_{x_1 \in H_1} \Phi(x_1, x_2).$$

PROOF OF THEOREM 3.6.1. We shall prove the last equality in case A is a compact set K. The general case follows by the capacitability argument. In fact, for an open set U we have $c_{k,p}^*(U)^p = c_{k,p}(U)^p = C_{k,p}(U)$ since $c_{k,p}$ is an inner capacity and an open set is $C_{k,p}$-capacitable. Since $c_{k,p}^*$ and $C_{k,p}$ are both outer capacities, it follows that $c_{k,p}^*(A)^p = C_{k,p}(A)$ for any set A. The capacitability theorem implies that analytic sets are $c_{k,p}^*$-capacitable. Hence $c_{k,p}(A) = c_{k,p}^*(A)$ and so $c_{k,p}(A)^p = C_{k,p}(A)$ for analytic sets A.

Let us rewrite $C_{k,p}(K)$ so that the mini-max theorem is applicable. First we claim

(3.6.1) $$C_{k,p}(K)^{-1/p} = \sup_{\substack{f \geq 0 \\ \|f\|_p \leq 1}} \inf_{x \in K} k(x, f).$$

Let α be the right hand side. Take an arbitrary $g \geq 0$ such that $k(\cdot, g) \geq 1$ on K. Then $f = g/\|g\|_p$ satisfies $\|f\|_p = 1$ and $k(\cdot, f) \geq 1/\|g\|_p$ on K. Hence $\alpha \geq 1/\|g\|_p$, and the arbitrariness of g yields

$$\alpha \geq C_{k,p}(K)^{-1/p}.$$

Conversely, take an arbitrary $f \geq 0$ such that $\|f\|_p \leq 1$ and let $m = \inf_{x \in K} k(x, f) > 0$. Then $g = f/m$ satisfies $k(\cdot, f) \geq 1$ on K. By definition

$$C_{k,p}(K)^{1/p} \leq \|g\|_p = \|f\|_p / m \leq 1/m$$

Thus $C_{k,p}(K)^{-1/p} \geq m$, whence the arbitrariness of f implies

$$C_{k,p}(K)^{-1/p} \geq \alpha.$$

Hence (3.6.1) follows.

Let

$$\mathfrak{M}_K^1 = \{\nu \geq 0 : \|\nu\|_1 = 1, \text{supp}(\nu) \subset K\}.$$

We claim next

(3.6.2) $$C_{k,p}(K)^{-1/p} = \sup_{\substack{f \geq 0 \\ \|f\|_p \leq 1}} \inf_{\nu \in \mathfrak{M}_K^1} k(\nu, f).$$

By ε_x we denote the point measure at x. Then $k(x, f) = k(\varepsilon_x, f)$. Since $\varepsilon_x \in \mathfrak{M}_K^1$ for $x \in K$, it follows from (3.6.1) the right hand side of (3.6.2) is not greater than $C_{k,p}(K)^{-1/p}$. The reverse inequality follows since

$$k(\nu, f) = \int_K k(x, f) d\nu(x) \geq \inf_{x \in K} k(x, f).$$

Thus we have (3.6.2).

Next we rewrite $c_{k,p}(K)$ as follows:

$$(3.6.3) \qquad c_{k,p}(K)^{-1} = \inf_{\nu \in \mathfrak{M}_K^1} \sup_{\substack{f \geq 0 \\ \|f\|_p \leq 1}} k(\nu, f).$$

Let β be the right hand side. Take an arbitrary test measure $\mu \neq 0$ for K, i.e. $\text{supp}(\mu) \subset K$ and $\|k(\mu, \cdot)\|_{p'} \leq 1$. Let $\nu = \mu / \|\mu\|_1$. Then $\nu \in \mathfrak{M}_K^1$ and

$$\|k(\nu, \cdot)\|_{p'} \leq 1 / \|\mu\|_1 \, .$$

In view of Lemma 3.1.1 we have

$$(3.6.4) \qquad \|k(\nu, \cdot)\|_{p'} = \sup_{\substack{f \geq 0 \\ \|f\|_p \leq 1}} k(\nu, f).$$

Hence $\beta \leq 1 / \|\mu\|_1$ and so

$$\beta \leq c_{k,p}(K)^{-1}.$$

Conversely, take an arbitrary measure $\nu \in \mathfrak{M}_K^1$. Then $\mu = \nu / \|k(\nu, \cdot)\|_{p'}$ is a test measure for K. Hence by (3.6.4)

$$c_{k,p}(K) \geq \|\mu\|_1 = 1 / \|k(\nu, \cdot)\|_{p'} = \left(\sup_{\substack{f \geq 0 \\ \|f\|_p \leq 1}} k(\nu, f) \right)^{-1}.$$

Since ν is arbitrary, it follows that

$$c_{k,p}(K) \geq \left(\inf_{\nu \in \mathfrak{M}_K^1} \sup_{\substack{f \geq 0 \\ \|f\|_p \leq 1}} k(\nu, f) \right)^{-1} = \beta^{-1}.$$

Thus (3.6.3) follows.

Now let us apply Lemma 3.6.1 with $H_1 = \{f \geq 0 : \|f\|_p \leq 1\}$, $H_2 = \mathfrak{M}_K^1$ and $\Psi(\nu, f) = k(\nu, f)$. In view of (3.6.2) and (3.6.3) we have

$$(3.6.5) \quad C_{k,p}(K)^{-1/p} = \sup_{\substack{f \geq 0 \\ \|f\|_p \leq 1}} \inf_{\nu \in \mathfrak{M}_K^1} k(\nu, f) = \inf_{\nu \in \mathfrak{M}_K^1} \sup_{\substack{f \geq 0 \\ \|f\|_p \leq 1}} k(\nu, f) = c_{k,p}(K)^{-1}.$$

Thus the theorem is proved. $\square$

We can extend (3.6.5) for analytic sets.

COROLLARY 3.6.1. *Let A be an analytic set. Put*

$$\mathfrak{M}_A^1 = \{\nu \geq 0 : \|\nu\|_1 = 1, \ \nu \text{ is concentrated on } A\}.$$

Then

$$C_{k,p}(A)^{-1/p} = c_{k,p}(A)^{-1} = \sup_{\substack{f \geq 0 \\ \|f\|_p = 1}} \inf_{\substack{\nu \in \mathfrak{M}_A^1 \\ \|k(\nu, \cdot)\|_{p'} < \infty}} k(\nu, f) = \inf_{\substack{\nu \in \mathfrak{M}_A^1 \\ \|k(\nu, \cdot)\|_{p'} < \infty}} \sup_{\substack{f \geq 0 \\ \|f\|_p = 1}} k(\nu, f).$$

PROOF. There are slight differences from the proof of (3.6.5). By the homogeneity we can replace the condition $\|f\|_p \leq 1$ by $\|f\|_p = 1$. We can put the additional condition $\|k(\nu, \cdot)\|_{p'} < \infty$ by the following reasoning: Suppose $\|k(\nu, \cdot)\|_{p'} = \infty$. Then $k(\nu, f) = \infty$ for $f \not\equiv 0$. Hence there is no effect on $\sup_f \inf_\nu$ and $\inf_\nu \sup_f$ if we remove ν with $\|k(\nu, \cdot)\|_{p'} = \infty$ from the infimum. $\square$

3.7. Relationship between capacitary distributions. Let A be a universally measurable set such that $0 < C_{k,p}(A) < \infty$. We have defined the $C_{k,p}$ and the $c_{k,p}$-capacitary distributions. Namely, $f \geq 0$ is the $C_{k,p}$-capacitary distribution if

$$\|f\|_p^p = C_{k,p}(A),$$
$$k(\cdot, f) \geq 1 \ C_{k,p}\text{-a.e. on } A;$$

and $\nu \geq 0$ is the $c_{k,p}$-capacitary distribution if

$$\nu(A^c) = 0, \ \|k(\nu, \cdot)\|_{p'} \leq 1,$$
$$\|\nu\|_1 = c_{k,p}(A).$$

The potentials $k(\cdot, f)$ and $k(\nu, \cdot)$ are called the $C_{k,p}$-capacitary potential and the $c_{k,p}$-capacitary potential, respectively. We are interested in the relationship among f, ν, $k(\cdot, f)$, $k(\nu, \cdot)$ and the *saddle point* $(\overline{\gamma}, \overline{g})$ for

$$(3.7.1) \qquad \sup_{\substack{g \geq 0 \\ \|g\|_p \leq 1}} \inf_{\substack{\gamma \in \mathfrak{M}_A^1 \\ \|k(\gamma, \cdot)\|_{p'} < \infty}} k(\gamma, g) = \inf_{\substack{\gamma \in \mathfrak{M}_A^1 \\ \|k(\gamma, \cdot)\|_{p'} < \infty}} \sup_{\substack{g \geq 0 \\ \|g\|_p \leq 1}} k(\gamma, g).$$

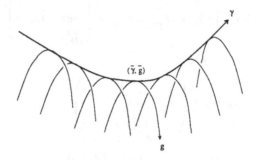

FIGURE 3.1

Here we say that $(\overline{\gamma}, \overline{g})$ is the saddle point if

$$(3.7.2) \qquad k(\overline{\gamma}, g) \leq k(\overline{\gamma}, \overline{g}) \leq k(\gamma, \overline{g})$$

for all $g \geq 0$ with $\|g\|_p \leq 1$ and all $\gamma \in \mathfrak{M}_A^1$ with $\|k(\gamma, \cdot)\|_{p'} < \infty$.
Let us begin with the following lemma.

LEMMA 3.7.1. *Let $\nu \geq 0$ and $\|k(\nu, \cdot)\|_{p'} < \infty$. Then*

$$\nu(E) = 0 \ \text{for every universally measurable set with } C_{k,p}(E) = 0.$$

PROOF. Take $\varepsilon > 0$. By definition there is $f \geq 0$ such that

$$k(\cdot, f) \geq 1 \text{ on } E, \ \|f\|_p < \varepsilon.$$

Observe that

$$\nu(E) \leq \int_E k(x, f) d\nu(x) \leq k(\nu, f) \leq \|f\|_p \|k(\nu, \cdot)\|_{p'} < \varepsilon \|k(\nu, \cdot)\|_{p'}.$$

Since $\varepsilon > 0$ is arbitrary, $\nu(E) = 0$. The lemma is proved. $\square$

THEOREM 3.7.1. *Let A be an analytic set such that $0 < C_{k,p}(A) < \infty$. Let f and ν be the $C_{k,p}$-capacitary distribution and the $c_{k,p}$-capacitary distribution for A, respectively. Then*

(3.7.3)
$$f(y)^{p-1} = c_{k,p}(A)^{p-1}k(\nu, y) \text{ for a.e. } y,$$
ν *is concentrated on B, where*
$$B = \{x : k(x, f) = 1\} \cap A, \; C_{k,p}(A) = C_{k,p}(B).$$

Moreover, (3.7.1) has a saddle point $(\overline{\gamma}, \overline{g})$ and we have

(3.7.4)
$$k(\overline{\gamma}, \overline{g}) = c_{k,p}(A)^{-1},$$
$$f = c_{k,p}(A)\overline{g},$$
$$\nu = c_{k,p}(A)\overline{\gamma}.$$

PROOF. Let us begin by proving (3.7.4). Let f and ν be the $C_{k,p}$-capacitary distribution and the $c_{k,p}$-capacitary distribution for A, respectively. Put $\overline{g} = c_{k,p}(A)^{-1}f$ and $\overline{\gamma} = c_{k,p}(A)^{-1}\nu$. Observe from Theorem 3.6.1 that

$$\|\overline{g}\|_p = c_{k,p}(A)^{-1} \|f\|_p = c_{k,p}(A)^{-1}C_{k,p}(A)^{1/p} = 1;$$

$\overline{\gamma}$ is concentrated on A and

$$\|\overline{\gamma}\|_1 = c_{k,p}(A)^{-1} \|\nu\|_1 = c_{k,p}(A)^{-1}c_{k,p}(A) = 1,$$
$$\|k(\overline{\gamma}, \cdot)\|_{p'} = c_{k,p}(A)^{-1} \|k(\nu, \cdot)\|_{p'} \le c_{k,p}(A)^{-1} < \infty.$$

Thus $\|\overline{g}\|_p = 1$, $\overline{\gamma} \in \mathfrak{M}_A^1$ and $\|k(\overline{\gamma}, \cdot)\|_{p'} < \infty$. Moreover, observe that

$$k(\cdot, \overline{g}) = c_{k,p}(A)^{-1}k(\cdot, f) \ge c_{k,p}(A)^{-1} \; C_{k,p}\text{-a.e. on } A.$$

Take $\gamma \in \mathfrak{M}_A^1$ such that $\|k(\gamma, \cdot)\|_{p'} < \infty$. Then Lemma 3.7.1 says that $k(\cdot, \overline{g}) \ge c_{k,p}(A)^{-1}$ γ-a.e. on A. Hence

$$k(\gamma, \overline{g}) = \int_A k(x, \overline{g})d\gamma(x) \ge c_{k,p}(A)^{-1}.$$

On the other hand, for any $g \ge 0$ with $\|g\|_p = 1$ we have

$$k(\overline{\gamma}, g) \le \|g\|_p \|k(\overline{\gamma}, \cdot)\|_{p'} \le c_{k,p}(A)^{-1}.$$

In particular, we have $k(\overline{\gamma}, \overline{g}) = c_{k,p}(A)^{-1}$, whence (3.7.2) holds. Thus $(\overline{\gamma}, \overline{g})$ is a saddle point and (3.7.4) holds.

Now we show (3.7.3). Using (3.7.4), we obtain,

(3.7.5)
$$k(\nu, f) = k(c_{k,p}(A)\overline{\gamma}, c_{k,p}(A)\overline{g}) = c_{k,p}(A)^2 k(\overline{\gamma}, \overline{g}) = c_{k,p}(A) = \|f\|_p = \|\nu\|_1.$$

Hölder's inequality yields

$$\|f\|_p = k(\nu, f) \le \|k(\nu, \cdot)\|_{p'} \|f\|_p \le 1 \cdot \|f\|_p.$$

Thus the equality must hold in Hölder's inequality. Hence

$$tk(\nu, y) = f(y)^{p-1} \text{ for a.e. } y$$

with some constant t. By an calculation it turns out that $t = \|f\|_p^{p-1} = c_{k,p}(A)^{p-1}$. Thus $f(y)^{p-1} = c_{k,p}(A)^{p-1}k(\nu, y)$ for a.e. y. Observe that ν is concentrated on A and

$k(\cdot, f) \geq 1$ $C_{k,p}$-a.e. on A. Since $\|k(\nu, \cdot)\|_{p'} \leq 1 < \infty$, it follows from Lemma 3.7.1 that $k(\cdot, f) \geq 1$ ν-a.e. Hence

$$\|\nu\|_1 \leq \int_A k(x, f) d\nu(x) = k(\nu, f) = \|\nu\|_1$$

by (3.7.5). The first inequality must be an equality and so

$$\nu(\{x \in A : k(x, f) > 1\}) = 0,$$

or ν is concentrated on $B = \{x \in A : k(x, f) = 1\}$. By definition $c_{k,p}(B) \geq \|\nu\|_1 = c_{k,p}(A)$. But B is a subset of A, and hence $c_{k,p}(B) = c_{k,p}(A)$. Thus (3.7.3) is proved. The theorem follows. $\square$

Remark. Suppose that there is a saddle point $(\overline{\gamma}, \overline{g})$. Then we can show that there are capacitary distributions f and ν. In view of (3.7.2) and a trivial inequality

$$\sup_{\substack{g \geq 0 \\ \|g\|_p \leq 1}} \inf_{\substack{\gamma \in \mathfrak{M}_A^1 \\ \|k(\gamma, \cdot)\|_{p'} < \infty}} k(\gamma, g) \leq \inf_{\substack{\gamma \in \mathfrak{M}_A^1 \\ \|k(\gamma, \cdot)\|_{p'} < \infty}} \sup_{\substack{g \geq 0 \\ \|g\|_p \leq 1}} k(\gamma, g)$$

we have

$$k(\overline{\gamma}, \overline{g}) = \sup_{\substack{g \geq 0 \\ \|g\|_p \leq 1}} \inf_{\substack{\gamma \in \mathfrak{M}_A^1 \\ \|k(\gamma, \cdot)\|_{p'} < \infty}} k(\gamma, g) = \inf_{\substack{\gamma \in \mathfrak{M}_A^1 \\ \|k(\gamma, \cdot)\|_{p'} < \infty}} \sup_{\substack{g \geq 0 \\ \|g\|_p \leq 1}} k(\gamma, g).$$

Thus by Corollary 3.6.1, $k(\overline{\gamma}, \overline{g}) = c_{k,p}(A)^{-1}$. Since $k(\gamma, \overline{g}) \geq k(\overline{\gamma}, \overline{g}) \geq c_{k,p}(A)^{-1}$ for any $\gamma \in \mathfrak{M}_A^1$, it follows that

$$k(\cdot, \overline{g}) \geq c_{k,p}(A)^{-1} \text{ on } A \ C_{k,p}\text{-a.e}$$

Hence $f = c_{k,p}(A)\overline{g}$ is the $C_{k,p}$-capacitary distribution. On the other hand, $k(\overline{\gamma}, g) \leq k(\overline{\gamma}, \overline{g}) = c_{k,p}(A)^{-1}$ for any $g \geq 0$ with $\|g\|_p = 1$, which implies that

$$\|k(\overline{\gamma}, \cdot)\|_{p'} \leq c_{k,p}(A)^{-1}$$

by Lemma 3.1.1. Thus $\nu = c_{k,p}(A)\overline{\gamma}$ is a $c_{k,p}$-capacitary distribution.

3.8. Capacitary measures and capacitary potentials. For a measure ν we define the nonlinear potential by

$$V_{k,p}^\nu(x) = \int k(x, y)k(\nu, y)^{p'-1} dy,$$

where as usual $p' = p/(p-1)$. By Fubini's theorem we have

$$(3.8.1) \qquad \int V_{k,p}^\nu d\nu = \int k(\nu, y)^{p'} dy.$$

DEFINITION. Let A be an arbitrary set with $0 < C_{k,p}(A) < \infty$. We say that μ_A is the *capacitary measure* for A if $\operatorname{supp}(\mu_A) \subset \overline{A}$, $\mu_A(\overline{A}) = C_{k,p}(A)$,

$$V_{k,p}^{\mu_A} \geq 1 \quad C_{k,p}\text{-a.e. on } A \text{ and}$$
$$V_{k,p}^{\mu_A} \leq 1 \quad \text{on } \operatorname{supp}(\mu_A).$$

We say that $V_{k,p}^{\mu_A}$ is the *capacitary potential* for A.

In this subsection we shall show the existence of capacitary measures. For a compact set K the capacitary measure μ_K always exists. This fact can be easily proved by Theorem 3.7.1.

THEOREM 3.8.1. *Let K be a compact set with $0 < C_{k,p}(K) < \infty$. Then K has the capacitary measure μ_K.*

PROOF. In view of Theorem 3.4.1, there is the $C_{k,p}$-capacitary distribution f for K; in view of Theorem 3.5.2 there is the $c_{k,p}$-capacitary distribution ν for K. By Theorem 3.7.1 they are related as

$$f = c_{k,p}(K)k(\nu,\cdot)^{p'-1} \quad \text{a.e. on } \mathbb{R}^n,$$
$$k(\cdot,f) \geq 1 \quad C_{k,p}\text{-a.e. on } K,$$
$$k(\cdot,f) = 1 \quad \nu\text{-a.e.}$$

Let $\mu_K = c_{k,p}(K)^{p-1}\nu$. Then $\operatorname{supp}(\mu_K) = \operatorname{supp}(\nu) \subset K$, $\mu_K(K) = c_{k,p}(K)^{p-1}\nu(K) = c_{k,p}(K)^p = C_{k,p}(K)$, $f = k(\mu_K,\cdot)^{p'-1}$, $k(\cdot,f) = V_{k,p}^{\mu_K}$ and

$$V_{k,p}^{\mu_K} \geq 1 \quad C_{k,p}\text{-a.e. on } K,$$
$$V_{k,p}^{\mu_K} = 1 \quad \mu_K\text{-a.e.}$$

By the l.s.c. of $V_{k,p}^{\mu_K}$ we have $V_{k,p}^{\mu_K} \leq 1$ on $\operatorname{supp}(\mu_K)$. Thus μ_K is the capacitary measure for K. $\square$

For the existence of the capacitary measure of an arbitrary set we need an additional assumption on k.

THEOREM 3.8.2. *Assume that there is a dense subset $\mathcal{D}$ of $L_+^p(\mathbb{R}^n)$ such that $k(\cdot,\varphi)$ is continuous and $\lim_{|x|\to\infty} k(x,\varphi) = 0$ for $\varphi \in \mathcal{D}$. Let A be an arbitrary set with $0 < C_{k,p}(A) < \infty$. Then A has a capacitary measure μ_A.*

PROOF. It is easy to find G_δ-set H and K_σ-set S such that $A \subset H \subset \overline{A}$, $S \subset H$ and $C_{k,p}(A) = C_{k,p}(H) = C_{k,p}(S)$. In view of Corollary 3.4.1, we see that A, H and S have the same capacitary distribution f. Let K_j be compact sets with $K_j \uparrow S$ and let f_j and ν_j be the $C_{k,p}$-capacitary distribution and $c_{k,p}$-capacitary distribution for K_j, respectively. As observed in the proof of Theorem 3.8.1, we see that $\mu_{K_j} = c_{k,p}(K_j)^{p-1}\nu_j$ is the capacitary measure for K_j, $f_j = k(\mu_{K_j},\cdot)^{p'-1}$ and $k(\cdot,f_j) = V_{k,p}^{\mu_{K_j}}$. Observe that $\left\|\mu_{K_j}\right\|_1 = C_{k,p}(K_j) \uparrow C_{k,p}(S) = C_{k,p}(A) < \infty$. Hence, if necessary taking a subsequence, we may assume that μ_{K_j} converges vaguely to μ_A. We shall show that this μ_A is the capacitary measure for A.

First we claim $f = k(\mu_A,\cdot)^{p'-1}$. In fact, the vague convergence implies

$$\int k(\mu_{K_j},y)\varphi(y)dy = \int k(x,\varphi)d\mu_{K_j}(x) \to \int k(x,\varphi)d\mu_A(x) = \int k(\mu_A,y)\varphi(y)dy$$

for all $\varphi \in \mathcal{D}$. On the other hand $f_j \to f$ in $L^p(\mathbb{R}^n)$ (See Exercise 3.8.1) and hence $k(\mu_{K_j},\cdot) = f_j^{p-1} \to f^{p-1}$ in $L^{p'}(\mathbb{R}^n)$, so that

$$\int k(\mu_{K_j},y)\varphi(y)dy \to \int f(y)^{p-1}\varphi(y)dy$$

for all $\varphi \in \mathcal{D}$. Hence the denseness of $\mathcal{D}$ implies that $f^{p-1} = k(\mu_A,\cdot)$ or equivalently $f = k(\mu_A,\cdot)^{p'-1}$. As a result

$$V_{k,p}^{\mu_A} = k(\cdot,f) \geq 1 \quad C_{k,p}\text{-a.e. on } A.$$

It is obvious that $\operatorname{supp}(\mu_A) \subset \overline{A}$. We have also

(3.8.2) $$V_{k,p}^{\mu_A} \leq 1 \quad \text{on } \operatorname{supp}(\mu_A).$$

In fact, if $x \in \operatorname{supp}(\mu_A)$, then we find $x_j \in \operatorname{supp}(\mu_{K_j})$ such that $x_j \to x$. We have $V_{k,p}^{\mu_{K_j}}(x_j) = k(x_j,f_j) \leq 1$ and hence $V_{k,p}^{\mu_A}(x) = k(x,f) \leq 1$ by Lemma 3.3.2.

Finally, it suffices to show that $\|\mu_A\|_1 = C_{k,p}(A)$. Integrate both sides of (3.8.2) with respect to μ_A and use (3.8.1) to obtain

$$C_{k,p}(A) = \|f\|_p^p = \int k(\mu_A, y)^{p'} dy = \int V_{k,p}^{\mu_A} d\mu_A \le \|\mu_A\|_1.$$

On the other hand, the vague convergence implies that $\|\mu_A\|_1 \le \liminf_{j\to\infty} \|\mu_{K_j}\|_1 = C_{k,p}(A)$, and hence $\|\mu_A\|_1 = C_{k,p}(A)$. Thus, μ_A is the capacitary measure for A. The proof is complete. $\square$

Exercise 3.8.1. Observe $\|f_j\|_p \uparrow \|f\|_p$ and $f_j \uparrow f$ a.e. Using these, prove $\|f_j - f\|_p \to 0$.

Remark. The kernel k satisfies the assumptions in Theorem 3.8.2 if for each compact set K

$$\lim_{|x|\to\infty} \int_K k(x, y) dy = 0,$$

$$\lim_{z\to x} \int_K |k(z, y) - k(x, y)| dy = 0 \quad \text{for all } x \in \mathbb{R}^n.$$

Let $\kappa(t)$ be a nonnegative l.s.c. function for $t > 0$. If $\int_0 \kappa(t) t^{n-1} dt < \infty$ and $\lim_{t\to\infty} \kappa(t) = 0$, then $k(x, y) = \kappa(|x - y|)$ satisfies the above condition.

Remark. Let k satisfy the assumptions in Theorem 3.8.2 and furthermore that

(3.8.3) $k(\mu, \cdot) = k(\nu, \cdot) \implies \mu = \nu$

for nonnegative measures μ and ν. Then the capacitary measure is unique. In fact, let $0 < C_{k,p}(A) < \infty$ and μ_A the capacitary measure for A. Then $V_{k,p}^{\mu_A} = k(\cdot, f) \ge 1$ $C_{k,p}$-a.e. on A with $f = k(\mu_A, \cdot)^{p'-1}$. Moreover,

$$\|f\|_p^p = \int k(\mu_A, y)^{p'} dy = \int V_{k,p}^{\mu_A} d\mu_A \le \|\mu_A\|_1 = C_{k,p}(A).$$

Hence, f is the $C_{k,p}$-capacitary distribution for A. From the uniqueness of $C_{k,p}$-capacitary distribution, we have that $k(\mu_A, \cdot)$ is unique and hence so is the capacitary measure by (3.8.3).

4. Capacity of balls

4.1. Introduction. This part is taken from H. Aikawa [8]. In this section we consider a kernel of convolution type $k(x, y) = k(x - y)$. In this case

$$C_{k,p}(E) = \inf\{\|f\|_p^p : f \ge 0, k * f \ge 1 \text{ on } E\},$$

$$c_{k,p}(E) = \sup\{\|\nu\|_1 : \nu \text{ is concentrated on } E, \|k * \nu\|_{p'} \le 1\}.$$

We are interested in estimating the capacity of balls. By definition our capacity is translation invariant, i.e., $C_{k,p}(E + x) = C_{k,p}(E)$ for any $x \in \mathbb{R}^n$. Thus it is sufficient to consider the capacity of balls with center at the origin. For simplicity we let $B(r) = B(0, r)$.

Let us begin with the classical case. Let $k(t) \not\equiv 0$ be a nonnegative nonincreasing l.s.c. function on $t > 0$ such that $\lim_{t\to\infty} k(t) = 0$. With a slight abuse of notation, we write $k(x) = k(|x|)$ for $x \in \mathbb{R}^n$. Throughout this section we assume that k is locally integrable, i.e.

(4.1.1) $\int_0 k(t) t^{n-1} dt < \infty.$

For a compact set E we define a classical capacity by

$$C_k(E) = \sup\{\|\nu\|_1 : \nu \text{ is concentrated on } E, k * \nu \leq 1 \text{ on } \mathbb{R}^n\}.$$

By the minimax theorem we have the alternative definition

$$C_k(E) = \inf\{\|\mu\|_1 : k * \mu \geq 1 \text{ on } E\}.$$

In a standard fashion we can extend $C_k(E)$ to an arbitrary set E. Let

$$\overline{k}(r) = \frac{1}{r^n} \int_0^r k(t)t^{n-1}dt.$$

This is a multiple of the volume mean of k. The capacity of a ball can be estimated by $\overline{k}$.

By the symbol A we denote an absolute positive constant whose value is unimportant and may change from line to line. We shall say that two positive functions f and g are comparable, written $f \approx g$, if and only if there exists a constant A such that $A^{-1}g \leq f \leq Ag$. In this case the constant A is called the *constant of comparison*.

PROPOSITION 4.1.1.

$$C_k(B(r)) \approx \overline{k}(r)^{-1}.$$

PROOF. Let $d\nu = |B(r)|^{-1}\chi_{B(r)}dx$. Then $\|\nu\|_1 = 1$ and ν is concentrated on $B(r)$. Since k is nonincreasing, it follows that

$$k * \nu(x) = \frac{1}{|B(r)|} \int_{|y|<r} k(x-y)dy \leq \frac{1}{|B(r)|} \int_{|y|<r} k(y)dy = A_n\overline{k}(r)$$

for any $x \in \mathbb{R}^n$. Hence

$$C_k(B(r)) \geq \frac{\|\nu\|_1}{A_n\overline{k}(r)} = \frac{1}{A_n}\overline{k}(r)^{-1}$$

by the first definition of C_k. Observe also that if $x \in B(r)$, then

$$k * \nu(x) \geq \frac{1}{|B(r)|} \int_{S_x} k(y)dy,$$

where $S_x = B(r) \cap B(x,r)$. It is easy to see that S_x contains the intersection of $B(r)$ and a cone with vertex at the origin, axis along with the line segment $0x$ and aperture depending only on the dimension. Hence, by using a polar coordinate, we obtain

$$\int_{S_x} k(y)dy \geq A \int_{|y|<r} k(y)dy,$$

whence $k * \nu(x) \geq A\overline{k}(r)$. The second definition of C_k yields

$$C_k(B(r)) \leq \frac{\|\nu\|_1}{A_n\overline{k}(r)} = \frac{1}{A_n}\overline{k}(r)^{-1}.$$

Thus the proposition follows. $\square$

Following [8], we shall show estimates for L^p-capacity corresponding to the above proposition. The Kerman-Sawyer inequality, a norm inequality between a fractional maximal function with respect to $\overline{k}$ and the convolution $k * \mu$, will play an important role. In the next section we start with some estimates of $\overline{k}$. Then, the Kerman-Sawyer inequality will be stated and applied to estimate the capacity of balls in the following sections.

4.2. Preliminaries. As was observed in the introduction, the averaged kernel $\overline{k}$ plays an important role. In fact, $\overline{k}$ always satisfies the doubling condition, though the original kernel k itself may not satisfy it. Let us begin with

LEMMA 4.2.1. *$\overline{k}$ is a positive continuous kernel such that:*
(i) $n^{-1}k(r) \leq \overline{k}(r)$.
(ii) $\overline{k}$ is nonincreasing and satisfies the doubling condition, i.e.

$$\overline{k}(2r) \leq \overline{k}(r) \leq 2^n \overline{k}(2r).$$

PROOF. By definition we can easily prove the positivity and continuity of $\overline{k}$. For (i) observe that

$$\overline{k}(r) \geq \frac{1}{r^n} \int_0^r k(r) t^{n-1} dt = n^{-1}k(r).$$

For (ii) we take $a \geq 1$. Then

$$
\begin{aligned}
\overline{k}(ar) &= \frac{1}{(ar)^n} \int_0^{ar} k(t) t^{n-1} dt \\
&\leq \frac{1}{(ar)^n} \left(\int_0^r k(t) t^{n-1} dt + \int_r^{ar} k(r) t^{n-1} dt \right) \\
&= \frac{1}{(ar)^n} \left(r^n \overline{k}(r) + k(r) \frac{(ar)^n - r^n}{n} \right) \\
&\leq \frac{1}{(ar)^n} \left(r^n \overline{k}(r) + ((ar)^n - r^n) \overline{k}(r) \right) \\
&= \overline{k}(r)
\end{aligned}
$$

Thus $\overline{k}$ is nonincreasing. Observe

$$\overline{k}(2r) = \frac{1}{(2r)^n} \int_0^{2r} k(t) t^{n-1} dt \geq 2^{-n} \frac{1}{r^n} \int_0^r k(t) t^{n-1} dt = 2^{-n} \overline{k}(r).$$

The lemma is proved. $\square$

We shall show that $\overline{k}$ is controlled by k in a certain sense. To this end we introduce the maximal function.

DEFINITION. Let f be a locally integrable function. We define

$$\mathcal{M}f(x) = \sup_{r>0} \frac{1}{|B(x,r)|} \int_{B(x,r)} |f| dy = \sup_{r>0} \frac{1}{|B(x,r)|} |f| * \chi_{B(0,r)}(x).$$

For a signed measure μ we define similarly

$$\mathcal{M}\mu(x) = \sup_{r>0} \frac{|\mu|(B(x,r))}{|B(x,r)|}.$$

The following result is well known (see e.g. [62]).

LEMMA 4.2.2. *(i) $\mathcal{M}f$ is an l.s.c. function.*
(ii) $|f(x)| \leq \mathcal{M}f(x)$ for a.e. x.
(iii) $f \mapsto \mathcal{M}f$ is of weak type (1,1); and of strong type (q,q) for $1 < q \leq \infty$ i.e.

$$\|\mathcal{M}f\|_q \leq A_q \|f\|_q.$$

LEMMA 4.2.3. *There is a positive constant A depending only on the dimension such that*

$$\overline{k}(x) \leq A\mathcal{M}k(x).$$

PROOF. By definition

$$\overline{k}(x) = \frac{1}{|x|^n} \int_0^{|x|} k(t)t^{n-1}dt = \frac{A}{|B(0,|x|)|} \int_{B(0,|x|)} k(y)dy,$$

and

$$\mathcal{M}k(x) \geq \frac{1}{|B(x,2|x|)|} \int_{B(x,2|x|)} k(y)dy.$$

Since $B(0,|x|) \subset B(x,2|x|)$, we have the required inequality. $\square$

In view of Lemma 4.2.2 and Lemma 4.2.3 we have for $1 < q \leq \infty$,

$$(4.2.1) \qquad \left\|\overline{k}\right\|_q \leq A \left\|\mathcal{M}k\right\|_q \leq A \left\|k\right\|_q,$$

where A depends only on n and q. However, we shall assume later that $\|k\|_{p'} = \infty$. In this case the above estimate is irrelevant. We have instead

LEMMA 4.2.4. *Let $1 < q < \infty$. Then, for all $r > 0$,*

$$\int_r^\infty \overline{k}(t)^q t^{n-1}dt \leq Ar^n\overline{k}(r)^q + A \int_r^\infty k(t)^q t^{n-1}dt,$$

where A depends only on n and q.

PROOF. Let $r > 0$ and let

$$k_r(t) = \min\{k(r), k(t)\}.$$

Applying (4.2.1), we obtain

$$(4.2.2) \qquad \left\|\overline{k}_r\right\|_q \leq A \left\|k_r\right\|_q.$$

Let $t > 2r$. Then

$$\begin{aligned}
\overline{k}_r(t) &= \frac{1}{t^n} \left(\int_0^r k(r)\tau^{n-1}d\tau + \int_r^t k(\tau)\tau^{n-1}d\tau \right) \\
&= \frac{1}{t^n} \left(\int_0^r (k(r) - k(\tau))\tau^{n-1}d\tau + \int_0^t k(\tau)\tau^{n-1}d\tau \right) \\
&\geq -\frac{r^n}{t^n}\overline{k}(r) + \overline{k}(t).
\end{aligned}$$

Hence (4.2.2) yields

$$\begin{aligned}
\int_{2r}^\infty \overline{k}(t)^q t^{n-1}dt &\leq A \left(\int_0^r k(r)^q t^{n-1}dt + \int_r^\infty k(t)^q t^{n-1}dt \right) + A \int_{2r}^\infty \left(\frac{r^n}{t^n}\overline{k}(r) \right)^q t^{n-1}dt \\
&= A \left(r^n k(r)^q + \int_r^\infty k(t)^q t^{n-1}dt \right) + Ar^n\overline{k}(r)^q \\
&\leq Ar^n\overline{k}(r)^q + A \int_r^\infty k(t)^q t^{n-1}dt.
\end{aligned}$$

It follows from Lemma 4.2.1 (ii) that

$$\int_r^\infty \overline{k}(t)^q t^{n-1}dt \leq A \int_{2r}^\infty \overline{k}(t)^q t^{n-1}dt.$$

Thus we have the required inequality. $\square$

Remark. In general, $k(t)$ and $\overline{k}(t)$ behave differently at 0. In fact if

$$k(t) \approx \frac{1}{t^n} \frac{1}{\left(\log \frac{1}{t}\right)^2} \text{ as } t \to 0,$$

then

$$\overline{k}(t) \approx \frac{1}{t^n} \frac{1}{\log \frac{1}{t}} \text{ as } t \to 0,$$

so that $\overline{k}$ is not locally integrable.

4.3. Kerman-Sawyer inequality. In this section we introduce a norm inequality between a fractional maximal function with respect to $\overline{k}$ and the convolution $k * \mu$ which is due to Kerman-Sawyer [39]. Their inequality is very useful to study L^p capacity for a general kernel (including the Riesz and the Bessel kernels). We define a generalized maximal function with respect to k by

$$M_k\mu(x) = \sup_{r>0} \overline{k}(r)\mu(B(x,r)).$$

If μ has the density f with respect to the Lebesgue measure, we write $M_k f$ for $M_k\mu$. Kerman-Sawyer [39, (2,7) and (2.8)] proved

THEOREM 4.3.1. *Let* $1 < q < \infty$. *Then, for any measure*

$$A^{-1} \|M_k\mu\|_q \le \|k * \mu\|_q \le A \|M_k\mu\|_q,$$

where $A > 1$ *depends only on* n *and* q.

For the sake of the convenience, we include a proof of Theorem 4.3.1.

LEMMA 4.3.1.

$$M_k\mu(x) \le A\mathcal{M}(k * \mu)(x).$$

In view of Lemma 4.2.2 and Lemma 4.3.1 we have one half of Theorem 4.3.1.

COROLLARY 4.3.1. *Let* $1 < q < \infty$. *Then*

$$\|M_k\mu\|_q \le A \|\mathcal{M}(k * \mu)\|_q \le A \|k * \mu\|_q.$$

PROOF OF LEMMA 4.3.1. Let $x \in \mathbb{R}^n$ and let $r > 0$. For simplicity write $B = B(x,r)$ and $B^* = B(x,2r)$. We have

$$\int_{B^*} k * \mu \, dy \ge \int_{B^*} dy \int_B k(y - z) d\mu(z)$$

$$\ge \int_B d\mu(z) \int_{|y-z|<r} k(y - z) dy$$

$$\ge A r^n \overline{k}(r)\mu(B).$$

By definition

$$\mathcal{M}(k * \mu)(x) \ge \frac{1}{|B^*|} \int_{B^*} k * \mu \, dy \ge A\overline{k}(r)\mu(B(x,r)).$$

Since $r > 0$ is arbitrary, we have the required inequality. The lemma follows. $\square$

For the opposite estimate in Theorem 4.3.1 we need the following *good λ inequality.*

LEMMA 4.3.2. *There is a constant* $\gamma > 1$ *such that for all* $\lambda > 0$ *and all* β, $0 < \beta \le 1$

$$(4.3.1) \quad |\{x : k * \mu(x) > \gamma\lambda, M_k\mu(x) \le \beta\lambda\}| \le A\frac{\beta}{\gamma}|\{x : \mathcal{M}(k * \mu)(x) > \lambda\}|.$$

We start with the following elementary lemma.

LEMMA 4.3.3. *Let $r > 0$. If $|x| \geq r$, then*

$$k(x) \leq \frac{A}{r^n} \int_{B(x,r)} k(y)dy = \frac{A}{r^n} k * \chi_{B(0,r)}(x).$$

PROOF. Let $S = B(x,r) \cap B(0,|x|)$. Observe that $|S| \geq Ar^n$ and $k(y) \geq k(x)$ for $y \in S$. Hence the right hand side of the required inequality is not smaller than

$$\frac{A}{r^n} \int_S k(x)dy \geq Ak(x).$$

The lemma follows. □

PROOF OF LEMMA 4.3.2. Let us consider the right hand side of (4.3.1). By the lower semicontinuity

$$\Omega = \{x : \mathcal{M}(k * \mu)(x) > \lambda\}$$

is an open set. Let $\bigcup Q_j = \Omega$ be a *Whitney decomposition* of Ω (see [62, p.16]). Observe that

$$\mathrm{diam}(Q_j) \approx \mathrm{dist}(Q_j, \Omega^c).$$

Hence there is a constant $A > 1$ such that the cube $\tilde{Q}_j$ with the same center as Q_j but expanded A times from Q_j satisfies

$$(4.3.2) \qquad \tilde{Q}_j \cap \Omega^c \neq \emptyset.$$

We are going to estimate of the measure of the set

$$\{x \in Q_j : k * \mu(x) > \gamma\lambda, M_k\mu(x) \leq \beta\lambda\}$$

and to sum over j to obtain the required estimate. In the estimate, we consider only cubes Q_j that have a point x_j such that

$$(4.3.3) \qquad M_k\mu(x_j) \leq \beta\lambda.$$

For simplicity let us write Q and $\tilde{Q}$ for Q_j and $\tilde{Q}_j$. In view of (4.3.2) and (4.3.3) we estimate

$$|\{x \in Q : k * \mu(x) > \gamma\lambda\}|$$

under the condition

$(4.3.4) \qquad$ there is a point $x' \in \tilde{Q}$ such that $\mathcal{M}(k * \mu)(x') \leq \lambda$,

$(4.3.5) \qquad$ there is a point $x'' \in Q$ such that $M_k\mu(x'') \leq \beta\lambda$.

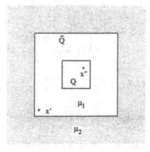

FIGURE 4.1

Let us write $\mu_1 = \mu|_{\widetilde{Q}}$ and $\mu_2 = \mu - \mu_1$. We claim that there is a constant A_0 depending only on n such that

$$(4.3.6) \qquad\qquad k * \mu_2(x) \leq A_0 \lambda \text{ for } x \in Q.$$

Let $x \in Q$ and let $r = \text{dist}(Q, \widetilde{Q}^c)$. From Lemma 4.3.3 we observe that

$$k * \mu_2(x) = \int_{y \notin \widetilde{Q}} k(x - y) d\mu(y) \leq \frac{A}{r^n} \int_{y \notin \widetilde{Q}} k * \chi_{B(0,r)}(x - y) d\mu(y).$$

Hence, if we choose A_1 such that $B(x, r) \subset B(x', A_1 r)$, then

$$k * \mu_2(x) \leq \frac{A}{r^n}(k * \mu_2) * \chi_{B(0,r)}(x) = \frac{A}{r^n} \int_{B(x,r)} k * \mu_2(y) dy \leq \frac{A}{r^n} \int_{B(x',A_1 r)} k * \mu_2(y) dy.$$

Therefore, by (4.3.4)

$$k * \mu_2(x) \leq A\mathcal{M}(k * \mu_2)(x') \leq A\lambda.$$

Thus (4.3.6) holds.

Let $\gamma > 2A_0$. Then

$$(4.3.7) \qquad \{x \in Q : k * \mu(x) > \gamma\lambda\} \subset \{x \in Q : k * \mu_1(x) > \frac{\gamma}{2}\lambda\}.$$

Let $R = \text{diam}(Q)$. Observe from Lemma 4.2.1 and (4.3.5) that

$$\int_Q k * \mu_1(x) dx = \int_{\widetilde{Q}} d\mu(y) \int_Q k(x - y) dx$$

$$\leq \int_{B(x'',AR)} d\mu(y) \int_{B(0,AR)} k(x) dx$$

$$\leq A\mu(B(x'', AR))\overline{k}(AR)R^n \leq AM_k\mu(x'')|Q| \leq A\beta\lambda|Q|.$$

Thus

$$\frac{\gamma}{2}\lambda \cdot |\{x \in Q : k * \mu_1(x) > \frac{\gamma}{2}\lambda\}| \leq \int_Q k * \mu_1(x) dx \leq A\beta\lambda|Q|.$$

Hence by (4.3.7)

$$|\{x \in Q : k * \mu(x) > \gamma\lambda\}| \leq A\frac{\beta}{\gamma}|Q|.$$

Collecting over all cubes, we obtain (4.3.1). The lemma is proved. $\square$

PROOF OF THE SECOND INEQUALITY OF THEOREM 4.3.1. By approximation, we may assume that $\|k * \mu\|_q < \infty$. Observe from Lemma 4.3.2 that

$$\|k * \mu\|_q^q = \int_0^\infty |\{x : k * \mu(x) > \gamma\lambda\}| d(\gamma\lambda)^q$$

$$\leq \gamma^q \int_0^\infty (|\{x : M_k\mu(x) > \beta\lambda\}| + |\{x : k * \mu(x) > \gamma\lambda, M_k\mu(x) \leq \beta\lambda\}|) d\lambda^q$$

$$\leq \gamma^q \int_0^\infty (|\{x : M_k\mu(x) > \beta\lambda\}| + A\frac{\beta}{\gamma}|\{x : \mathcal{M}(k * \mu)(x) > \lambda\}|) d\lambda^q$$

$$= \gamma^q \beta^{-q} \|M_k\mu\|_q^q + A\frac{\beta}{\gamma} \|\mathcal{M}(k * \mu)\|_q^q$$

$$\leq \gamma^q \beta^{-q} \|M_k\mu\|_q^q + A_2\frac{\beta}{\gamma} \|k * \mu\|_q^q.$$

We choose $\beta > 0$ so small that $A_2\beta/\gamma < 1/2$ to obtain the required norm inequality. The theorem is proved. $\square$

4.4. Capacity of balls. First of all we give two essential assumptions on the kernel k. Throughout this section we assume that

$$(4.4.1) \qquad \int_r^\infty k(t)^{p'} t^{n-1} dt < \infty \text{ for } r > 0,$$

$$(4.4.2) \qquad \int_0 k(t)^{p'} t^{n-1} dt = \infty.$$

These two assumptions are essential as we see in the following two lemmas. For the proof of Lemma 4.4.1 we use Lemma 4.2.4. Lemma 4.4.2 is easy.

LEMMA 4.4.1. *The following statements are equivalent:*
  (i) *(4.4.1) holds.*
  (ii) $\int_r^\infty \overline{k}(t)^{p'} t^{n-1} dt < \infty$ *for $r > 0$.*
  (iii) *There is a set E such that $C_{k,p}(E) > 0$.*
  (iv) *Every non-empty open set has positive $C_{k,p}$-capacity.*

LEMMA 4.4.2. *A point has vanishing $C_{k,p}$-capacity if and only if $k \notin L^{p'}(\mathbb{R}^n)$. In fact $c_{k,p}(\{0\}) = 1/\|k\|_{p'}$.*

The next theorem is our main result in this section.

THEOREM 4.4.1. *Let us assume (4.4.1) and (4.4.2). Then*

$$C_{k,p}(B(r)) \approx \left(\int_r^\infty \overline{k}(t)^{p'} t^{n-1} dt \right)^{1-p}.$$

PROOF. Take an arbitrary measure μ concentrated on $B(r)$ such that $\|k * \mu\|_{p'} \leq 1$. Then by Theorem 4.3.1 we have $\|M_k \mu\|_{p'} \leq A$. Since $B(r) \subset B(x, 2|x|)$ for $|x| > r$, it follows from Lemma 4.2.1 that

$$M_k \mu(x) \geq \overline{k}(2|x|) \|\mu\|_1 \geq A \overline{k}(|x|) \|\mu\|_1$$

for $|x| > r$. Hence

$$\int_{|x|>r} \left(\overline{k}(|x|) \|\mu\|_1 \right)^{p'} dx \leq A.$$

This implies

$$\|\mu\|_1 \leq A \left(\int_r^\infty \overline{k}(t)^{p'} t^{n-1} dt \right)^{-1/p'}.$$

Therefore Theorem 3.6.1 yields

$$C_{k,p}(B(r)) \leq A \left(\int_r^\infty \overline{k}(t)^{p'} t^{n-1} dt \right)^{-p/p'}.$$

Thus the one half of the estimate follows.
  For the opposite estimate we let $d\mu = \chi_{B(r)} dx$. We put

$$k_1(x) = \begin{cases} k(x) & \text{if } |x| < 3r, \\ 0 & \text{if } |x| \geq 3r \end{cases}$$

and put $k_2(x) = k(x) - k_1(x)$. Observe that $k_2 * \mu(x) = 0$ for $|x| \le 2r$ and $k_2 * \mu(x) \le \|\mu\|_1 k(\frac{|x|}{2})$ for $|x| > 2r$. Hence

$$\|k_2 * \mu\|_{p'}^{p'} \le A \|\mu\|_1^{p'} \int_{2r}^{\infty} k(\frac{t}{2})^{p'} t^{n-1} dt$$

$$\le A r^{np'} \int_{2r}^{\infty} \overline{k}(\frac{t}{2})^{p'} t^{n-1} dt$$

$$\le A r^{np'} \int_{r}^{\infty} \overline{k}(t)^{p'} t^{n-1} dt$$

by Lemma 4.2.1. On the other hand

$$\|k_1 * \mu\|_{p'} = \|k_1 * \chi_{B(r)}\|_{p'}$$

$$\le \|k_1\|_1 \|\chi_{B(r)}\|_{p'} = A r^n \overline{k}(r) r^{n/p'}$$

$$\le A r^n \left(\int_{r}^{2r} \overline{k}(t)^{p'} t^{n-1} dt \right)^{1/p'}$$

$$\le A r^n \left(\int_{r}^{\infty} \overline{k}(t)^{p'} t^{n-1} dt \right)^{1/p'}.$$

Thus

$$\|k * \mu\|_{p'} \le A r^n \left(\int_{r}^{\infty} \overline{k}(t)^{p'} t^{n-1} dt \right)^{1/p'}.$$

Hence

$$\tilde{\mu} = \left[r^n \left(\int_{r}^{\infty} \overline{k}(t)^{p'} t^{n-1} dt \right)^{1/p'} \right]^{-1} \mu$$

satisfies $\|k * \tilde{\mu}\|_{p'} \le A$. Theorem 3.6.1 yields

$$C_{k,p}(B(r)) \ge A \|\tilde{\mu}\|_1^p = A \left(\int_{r}^{\infty} \overline{k}(t)^{p'} t^{n-1} dt \right)^{1-p}.$$

The theorem follows. $\square$

4.5. Metric Property of Capacity. The estimate in Theorem 4.4.1 yields a comparison between capacity and Hausdorff measure. Let $h(r)$ be a positive nondecreasing function for $r > 0$. We put

$$H_{h,R}(E) = \inf\{\sum_j h(r_j) : E \subset \bigcup_j B(x_j, r_j), 0 < r_j < R\},$$

$$H_h(E) = \lim_{R \to 0} H_{h,R}(E).$$

The quantity $H_h(E)$ is called the Hausdorff measure of E with respect to the measure function h.

THEOREM 4.5.1. *Let* $h(r) = C_{k,p}(B(r))$. *Then,* $H_h(E) < \infty$ *implies* $C_{k,p}(E) = 0$.

Remark. It is immediate from the countable subadditivity that $H_h(E) = 0$ implies $C_{k,p}(E) = 0$. The above theorem means that the weak assumption $H_h(E) < \infty$ is sufficient for $C_{k,p}(E) = 0$.

PROOF. We shall prove the theorem by contradiction. Suppose that there is a set E such that $H_h(E) < \infty$ and $C_{k,p}(E) > 0$. Without loss of generality we may assume that E is compact. By an argument in [53, Lemma 9 and (40)] (See also Remark below) we can find a positive kernel $\tilde{k}$ satisfying (4.1.1) such that $k(r) \leq \tilde{k}(r)$, $C_{\tilde{k},p}(E) > 0$ and

$$\lim_{r \to 0} \frac{\tilde{k}(r)}{k(r)} = \infty.$$

We invoke Theorem 4.4.1 for the kernels k and $\tilde{k}$ to obtain

$$\lim_{r \to 0} \frac{\tilde{h}(r)}{h(r)} = 0,$$

where $\tilde{h}(r) = C_{\tilde{k},p}(B(r))$. Since $H_h(E) < \infty$, we conclude that $H_{\tilde{h}}(E) = 0$. The countable subadditivity yields $C_{\tilde{k},p}(E) = 0$, a contradiction. The theorem is proved. $\square$

Remark. The kernel $\tilde{k}(r)$ can be constructed in the following way. Since $C_{k,p}(E) > 0$, we have μ such that $\text{supp}(\mu) \subset E$ and $k * \mu \in L^{p'}(\mathbb{R}^n)$. Let

$$v_i(x) = \int_{2^{-i} \leq |x-y| < 2^{1-i}} k(x-y) d\mu(y).$$

Let us construct a sequence $\{a_i\}$ such that $a_i \geq 1$, $a_i \uparrow \infty$ and $\sum a_i v_i(x) \in L^{p'}(\mathbb{R}^n)$. If we have such a sequence, then the kernel $\tilde{k}(r)$ defined by $\tilde{k}(0) = \infty$,

$$\tilde{k}(r) = \begin{cases} a_i k(r) & \text{for } 2^{-i} \leq r < 2^{1-i}, i = 1, 2, \ldots \\ k(r) & \text{for } r \geq 1 \end{cases}$$

is the required kernel. The construction of $\{a_i\}$ is as follows. Since $\left\| \sum_{i=\ell}^{\infty} v_i \right\|_{p'} \to 0$ as $\ell \to \infty$, we have an increasing sequence $\{\ell_j\}$ such that $\sum_{j=1}^{\infty} \left\| \sum_{i=\ell_j}^{\infty} v_i \right\|_{p'} < \infty$. Then we can find $b_j \geq 1$, $b_j \uparrow \infty$ such that $\sum_{j=1}^{\infty} b_j \left\| \sum_{i=\ell_j}^{\infty} v_i \right\|_{p'} < \infty$ (See Exercise 4.5.1) Define $a_i = b_j$ for $\ell_j \leq i < \ell_{j+1}$. Then

$$\left\| \sum_{i=1}^{\infty} a_i v_i \right\|_{p'} \leq \sum_{j=1}^{\infty} \left\| \sum_{i=\ell_j}^{\ell_{j+1}-1} a_i v_i \right\|_{p'} \leq \sum_{j=1}^{\infty} b_j \left\| \sum_{i=\ell_j}^{\infty} v_i \right\|_{p'} < \infty.$$

Exercise 4.5.1. Let $\alpha_i > 0$ and $\sum_{i=1}^{\infty} \alpha_i < \infty$. Construct $\beta_i \geq 1$ such that $\beta_i \uparrow \infty$ and $\sum_{i=1}^{\infty} \alpha_i \beta_i < \infty$. (Hint. Consider ℓ_j such that $\sum_{i=\ell_j}^{\infty} \alpha_i < 4^{-j}$.)

We have an opposite comparison.

THEOREM 4.5.2. Suppose a measure function h satisfies

(4.5.1) $$\int_0 h(t)^{p'-1} \overline{k}(t)^{p'} t^{n-1} dt < \infty.$$

Then $C_{k,p}(E) = 0$ implies $H_h(E) = 0$.

PROOF. Without loss of generality we may assume that E is compact. Suppose $H_h(E) > 0$. Then by Frostman's lemma ([23, p.7]) we find a measure μ on E such that $0 < \mu(E) < \infty$ and $\mu(B(x,r)) \leq h(r)$ for $x \in \mathbb{R}^n$ and $r > 0$. Applying Lemma 5.2.1 (which is a corollary of the Kerman-Sawyer inequality), we obtain

$$\|k * \mu\|_{p'}^{p'} \leq A \|M_k \mu\|_{p'}^{p'} \leq A \int_{\mathbb{R}^n} d\mu(x) \int_0^{\infty} \overline{k}(t)^{p'} \mu(B(x,t))^{p'-1} t^{n-1} dt,$$

and
$$\|k * \mu\|_{p'}^{p'} \leq A \int_{\mathbb{R}^n} d\mu(x) \int_0^\infty \overline{k}(t)^{p'} (\min\{h(t), \|\mu\|_1\})^{p'-1} t^{n-1} dt.$$

By (4.5.1) and Lemma 4.2.4 the last integral is convergent and hence $C_{k,p}(E) > 0$ by Theorem 3.6.1. This is a contradiction. $\square$

We can show the sharpness of Theorem 4.5.1 and Theorem 4.5.2. For details we refer to [8].

Remark. We can easily see that the function $h(r) = C_{k,p}(B(r))$ does not satisfy (4.5.1). In general, the null sets for capacity cannot be characterized by those for Hausdorff measure.

5. Capacity under a Lipschitz mapping

5.1. Introduction. In this section we give an application of Theorem 4.3.1. Let
$$W_k^\mu = W_{k,q}^\mu = \int_0^\infty \overline{k}(t)^q \mu(B(x,t))^{q-1} t^{n-1} dt.$$

This kind of potential was used by Hedberg and Wolff [35] and is called sometimes a Wolff potential. The following is a generalization of their result.

THEOREM 5.1.1. *Suppose one of the following conditions holds:*
  (i) $k(r) \approx \overline{k}(r)$ *for all* $r > 0$.
  (ii) $k(r) \approx \overline{k}(r)$ *for* $0 < r < r_0$ *and* $\int_0^\infty k(t) t^{n-1} dt < \infty$.
Then
$$\|k * \mu\|_q^q \approx \int W_k^\mu d\mu.$$

Moreover, if (ii) holds, then for $R > 0$
$$\|k * \mu\|_q^q \approx \int d\mu(x) \int_0^R k(t)^q \mu(B(x,t))^{q-1} t^{n-1} dt.$$

COROLLARY 5.1.1. *Let* $1 < p < \infty$ *and* $q = p/(p-1)$. *Suppose* k *satisfies (i) or (ii) of Theorem 5.1.1. Then*
$$C_{k,p}(E) \approx \sup\{\|\mu\|^p : \int W_k^\mu(x) d\mu(x) \leq 1, \ \mu \text{ is concentrated on } E\}$$
for analytic sets E.

We shall consider how the capacity changes if the set is mapped by a Lipschitz mapping. Let φ be a Lipschitz mapping such that
$$|\varphi(x) - \varphi(y)| \leq L|x - y|.$$

THEOREM 5.1.2. *Let* k *satisfy (i) or (ii) of Theorem 5.1.1. Suppose*

(5.1.1) $\overline{k}(Lt) \geq A_L \overline{k}(t)$ *for all* $t > 0$.

Then
$$C_{k,p}(\varphi(E)) \leq A L^{n(1-p)} A_L^{-p} C_{k,p}(E).$$

Let k_α and g_α be the Riesz and the Bessel kernels. The corresponding capacities are called the Riesz and the Bessel capacities and denoted by $R_{\alpha,p}$ and $B_{\alpha,p}$, respectively. It is easy to see that k_α satisfies (i) of Theorem 5.1.1; g_α satisfies (ii) of Theorem 5.1.1. Moreover, we observe that k_α and g_α satisfy (5.1.1) with $A_L = AL^{\alpha-n}$ and $A_L = A \min\{L^{\alpha-n}, L^{-n}\}$, respectively. We can apply Theorem 5.1.2 to $R_{\alpha,p}$ and $B_{\alpha,p}$.

COROLLARY 5.1.2. *Let $1 < p < n/\alpha$ and let $R_{\alpha,p}$ be the Riesz capacity. Then*

$$R_{\alpha,p}(\varphi(E)) \leq AL^{n-\alpha p}R_{\alpha,p}(E).$$

COROLLARY 5.1.3. *Let $1 < p \leq n/\alpha$ and let $B_{\alpha,p}$ be the Bessel capacity. Then*

$$B_{\alpha,p}(\varphi(E)) \leq A\max\{L^{n-\alpha p}, L^n\}B_{\alpha,p}(E).$$

The above corollaries are given by Adams and Hedberg [3, Theorem 5.2.1].

Remark. A mapping φ satisfying $|\varphi(x) - \varphi(y)| \leq |x - y|$ is called a contraction mapping. It is well known that a classical capacity decreases under a contraction mapping. The same is true for many capacities (see Fuglede [31]). Surprisingly, it is not known that L^p-capacity has this property or not unless $p = 2$. For one dimensional case or for a special contraction mapping this is true ([7]). Theorem 5.1.2 implies the contraction property is approximately valid up to a multiplicative constant. At least, if a set has vanishing L^p-capacity, then so does its image under a contraction mapping.

5.2. Proof of Theorem 5.1.1.

LEMMA 5.2.1. *Let $0 < R \leq +\infty$. Define*

$$M_{k,R}\mu(x) = \sup_{0<r<R} \overline{k}(r)\mu(B(x,r)).$$

Then

$$\|M_{k,R}\mu\|_q^q \leq A \int d\mu \int_0^{4R} \overline{k}(t)^q\mu(B(x,t))^{q-1}t^{n-1}dt.$$

where A depends only on q, n and k. In particular,

$$\|M_k\mu\|_q^q \leq A \int d\mu \int_0^\infty \overline{k}(t)^q\mu(B(x,t))^{q-1}t^{n-1}dt.$$

PROOF. The following proof is based on Adams [2]. Observe that

(5.2.1) $$\overline{k}(r) \approx \overline{k}(2r) \quad \text{for all } r > 0.$$

(Note that k itself may not satisfy the doubling condition.) Hence

$$\int_0^{2R} \overline{k}(t)^q\mu(B(x,t))^q t^{-1}dt \geq \int_r^{2r} \overline{k}(2r)^q\mu(B(x,t))^q t^{-1}dt \geq A\overline{k}(r)^q\mu(B(x,r))^q$$

for all $0 < r < R$. Hence by definition

$$(M_{k,R}\mu(x))^q \leq A \int_0^{2R} \overline{k}(t)^q\mu(B(x,t))^q t^{-1}dt,$$

whence

$$\int (M_{k,R}\mu(x))^q dx \leq A \int_0^{2R} \overline{k}(t)^q t^{-1}dt \int \mu(B(x,t))^q dx.$$

Note

$$\int \mu(B(x,t))^q dx = \int \mu(B(x,t))^{q-1}\left(\int \chi_{|x-y|<t}(y)d\mu(y)\right)dx$$

$$= \int d\mu(y) \int \chi_{|x-y|<t}(x)\mu(B(x,t))^{q-1}dx$$

$$\leq \int d\mu(y) \int \chi_{|x-y|<t}(x)\mu(B(y,2t))^{q-1}dx$$

$$= \int \mu(B(y,2t))^{q-1}d\mu(y) \int \chi_{|x-y|<t}(x)dx$$

$$= At^n \int \mu(B(y,2t))^{q-1}d\mu(y).$$

Substituting this estimate into the above inequality, we obtain

$$\int (M_{k,R}\mu(x))^q dx \le A \int_0^{2R} \overline{k}(t)^q t^{-1} dt \cdot t^n \int \mu(B(x,2t))^{q-1} d\mu(x)$$

$$= A \int d\mu(x) \int_0^{2R} \overline{k}(t)^q \mu(B(x,2t))^{q-1} t^{n-1} dt$$

$$= A \int d\mu(x) \int_0^{4R} \overline{k}(t/2)^q \mu(B(x,t))^{q-1} t^{n-1} dt$$

$$\le A \int d\mu(x) \int_0^{4R} \overline{k}(t)^q \mu(B(x,t))^{q-1} t^{n-1} dt.$$

Here we used (5.2.1) in the last inequality. The lemma follows.　□

LEMMA 5.2.2.

$$\|k * \mu\|_q^q \ge A \int d\mu(x) \int_0^\infty k(2t)^q \mu(B(x,t))^{q-1} t^{n-1} dt.$$

PROOF. We have

$$\|k * \mu\|_q^q = \int k * (k * \mu)^{q-1} d\mu.$$

Let us give a lower estimate of $k * (k * \mu)^{q-1}$. Using polar coordinates, we deduce that

$$k * (k * \mu)^{q-1}(x) = \int_0^\infty k(t) dt \int_{|x-y|=t} (k * \mu)^{q-1}(y) d\sigma(y).$$

and that

$$k * \mu(y) \ge k(2t)\mu(B(x,t)) \quad \text{for } |x-y| = t.$$

Hence

$$k * (k * \mu)^{q-1}(x) \ge \int_0^\infty k(t) dt \int_{|x-y|=t} [k(2t)\mu(B(x,t))]^{q-1} d\sigma(y)$$

$$= A \int_0^\infty k(t) t^{n-1} [k(2t)\mu(B(x,t))]^{q-1} dt \ge A \int_0^\infty k(2t)^q \mu(B(x,t))^{q-1} t^{n-1} dt.$$

The lemma follows.　□

PROOF OF THEOREM 5.1.1 IN CASE (I). In case (i) we have

$$k(t) \approx \overline{k}(t) \approx \overline{k}(2t) \approx k(2t) \quad \text{for all } t > 0.$$

Hence, in view of Theorem 4.3.1, Lemma 5.2.1 and Lemma 5.2.2, we have

$$\|k * \mu\|_q^q \approx \|M_k\mu\|_q^q \le A \int d\mu(x) \int_0^\infty k(t)^q \mu(B(x,t))^{q-1} t^{n-1} dt \le A\|k * \mu\|_q^q.$$

Thus the theorem follows.　□

In order to treat case (ii), we prove the following lemma.

LEMMA 5.2.3. Let $\int_0^\infty k(t) t^{n-1} dt < \infty$. Then for each $R > 0$ there is a positive constant A depending only on R and k such that

$$M_k\mu(x) \le AM(M_{k,R}\mu)(x) \quad \text{a.e. } x \in \mathbb{R}^n.$$

PROOF. Let $m < M_k\mu(x)$. By definition we can find $r = r(x)$ such that $m < \overline{k}(r)\mu(B(x,r))$. If $r < R$, then

$$(5.2.2) \qquad m < M_{k,R}\mu(x) \leq \mathcal{M}(M_k\mu)(x),$$

where the second inequality holds for a.e. x by the properties of the maximal function. Suppose $r \geq R$. In general $M_{k,R}\mu(y) \geq \overline{k}(R)\mu(B(y,R)) \geq \overline{k}(R)\mu(B(y_0, R/2))$ for $|y - y_0| < R/2$, whence

$$(5.2.3) \qquad \int_{B(y_0,R/2)} M_{k,R}\mu(y)dy \geq AR^n\overline{k}(R)\mu(B(y_0, R/2)) \text{ for every } y_0.$$

We can cover $B(x,r)$ by balls $B(y_j, R/2)$ so that

$$\bigcup_j B(y_j, R/2) \subset B(x, r+R) \subset B(x, 2r),$$

$$\sum_j \chi_{B(y_j,R/2)} \leq A.$$

Then, applying (5.2.3) for $y_0 = y_j$, we obtain

$$\mu(B(x,r)) \leq \sum_j \mu(B(y_j, R/2)) \leq \frac{A}{R^n\overline{k}(R)} \sum_j \int_{B(y_j,R/2)} M_{k,R}\mu(y)dy$$

$$\leq \frac{A}{R^n\overline{k}(R)} \int_{B(x,2r)} M_{k,R}\mu(y)dy.$$

Dividing both sides by r^n, we obtain

$$\frac{\mu(B(x,r))}{r^n} \leq \frac{A}{R^n\overline{k}(R)} \mathcal{M}(M_{k,R}\mu)(x).$$

Let $I_R = \int_0^R k(t)t^{n-1}dt$ and $I_\infty = \int_0^\infty k(t)t^{n-1}dt$. Then we have $R^n\overline{k}(R) = I_R$ and $\overline{k}(r) = r^{-n}\int_0^r k(t)t^{n-1}dt \leq r^{-n}I_\infty$, so that

$$m < \overline{k}(r)\mu(B(x,r)) \leq A\frac{I_\infty}{I_R}\mathcal{M}(M_{k,R}\mu)(x).$$

This, together with (5.2.2), yields the required estimate. $\square$

COROLLARY 5.2.1. *Let $\int_0^\infty k(t)t^{n-1}dt < \infty$. Then*

$$\|M_{k,R}\mu\|_q \approx \|M_k\mu\|_q,$$

where the constant of comparison depends only on k, q and R.

PROOF. It is obvious that $\|M_{k,R}\mu\|_q \leq \|M_k\mu\|_q$. By Lemma 5.2.3 and the maximal inequality

$$\|M_k\mu\|_q \leq A\|\mathcal{M}(M_{k,R}\mu)\|_q \leq A\|M_{k,R}\mu\|_q.$$

$\square$

PROOF OF THEOREM 5.1.1 IN CASE (II). Let $R_0 = r_0/4$. Then, by assumption, Theorem 4.3.1, Lemma 5.2.1, Lemma 5.2.2 and Corollary 5.2.1 yield

$$\|k * \mu\|_q^q \approx \|M_k \mu\|_q^q \approx \|M_{k,R_0} \mu\|_q^q$$
$$\leq A \int d\mu(x) \int_0^{r_0} \overline{k}(t)^q \mu(B(x,t))^{q-1} t^{n-1} dt$$
$$\leq A \int d\mu(x) \int_0^{r_0} k(t)^q \mu(B(x,t))^{q-1} t^{n-1} dt$$
$$\leq A \int d\mu(x) \int_0^{\infty} k(t)^q \mu(B(x,t))^{q-1} t^{n-1} dt$$
$$\leq A \|k * \mu\|_q^q.$$

The above chain of inequalities shows the theorem for $r_0 \leq R \leq \infty$. Moreover, the constant of comparison is independent of R. For $0 < R < r_0$ we can prove the theorem in the same fashion, by considering $M_{k,R/4}\mu$ in place of $M_{k,R_0}\mu$. Note that the constant of comparison depends on R in this case. The theorem follows. $\square$

5.3. Proof of Theorem 5.1.2. For the proof of Theorem 5.1.2 we need the following well known lemma. The proof is taken from [31, Lemma].

LEMMA 5.3.1. *Let E and E' be compact spaces and φ a continuous surjection from E to E'. Then for any (Radon) measure μ' on E' there exists a measure μ on E such that*

$$\int_{E'} f' d\mu' = \int_E (f' \circ \varphi) d\mu \quad \text{for all l.s.c. functions } f' \text{ on } E'.$$

PROOF. We define a functional $p(f)$ for $f \in C(E)$ by $p(f) = \|\mu'\| \max f(E)$. It is easy to see that $p(f + g) \leq p(f) + p(g)$ and $p(tf) = tp(f)$ for $t > 0$. Thus $p(f)$ is subadditive and positive homogeneous. Define a linear form λ on the vector subspace $\{f' \circ \varphi : f' \in C(E')\}$ of $C(E)$ by

$$\lambda(f' \circ \varphi) = \int_{E'} f' d\mu'.$$

This is well defined since φ is surjective. We have

$$\lambda(f' \circ \varphi) \leq \|\mu'\| \max f'(E') = \|\mu'\| \max(f' \circ \varphi)(E) = p(f' \circ \varphi).$$

By the Hahn-Banach theorem (see e.g. [69, p.102]) we can extend λ to a linear form μ on $C(E)$ with $\mu(f) \leq p(f)$ for all $f \in C(E)$. In particular, if $f \leq 0$, then $\mu(f) \leq 0$, so that μ is a nonnegative Radon measure on E. Since μ is an extension of λ, the required identity holds for $f' \in C(E')$, and hence for l.s.c. f' by the monotone convergence theorem. $\square$

PROOF OF THEOREM 5.1.2. We may assume that E is compact. For simplicity we let $E' = \varphi(E)$. By Corollary 5.1.1 we can find a measure μ' on E' such that

$$\|\mu'\|^p = C_{k,p}(E'),$$

$$\int W_k^{\mu'} d\mu' \leq A.$$

By Lemma 5.3.1 we find a measure μ on E such that

(5.3.1) $$\int_{E'} f'(x) d\mu'(x) = \int_E f'(\varphi(x)) d\mu(x)$$

for all l.s.c. functions f' on E'. In particular,

(5.3.2) $$\|\mu\|^p = \|\mu'\|^p = C_{k,p}(E').$$

Observe from (5.3.1) that for every $x \in E$

$$(5.3.3) \quad \mu'(B(\varphi(x), t)) = \int \chi_{B(\varphi(x),t)}(y) d\mu'(y) = \int \chi_{B(\varphi(x),t)}(\varphi(y)) d\mu(y)$$

$$\geq \int \chi_{B(x,t/L)}(y) d\mu(y) = \mu(B(x, t/L)),$$

where the inequality follows since $|x - y| < t/L$ implies $|\varphi(x) - \varphi(y)| < t$. Using (5.3.1) again, we obtain

$$\int_{E'} W_k^{\mu'}(x) d\mu'(x) = \int_E W_k^{\mu'}(\varphi(x)) d\mu(x) = \int_E d\mu(x) \int_0^\infty \overline{k}(t)^q \mu'(B(\varphi(x), t))^{q-1} t^{n-1} dt$$

$$\geq \int_E d\mu(x) \int_0^\infty \overline{k}(t)^q \mu(B(x, t/L))^{q-1} t^{n-1} dt = L^n \int_E d\mu(x) \int_0^\infty \overline{k}(Lt)^q \mu(B(x, t))^{q-1} t^{n-1} dt.$$

Here we use (5.3.3) and the change of the variable. By (5.1.1) we have

$$\int_{E'} W_k^{\mu'}(x) d\mu'(x) \geq L^n \int_E d\mu(x) \int_0^\infty [A_L \overline{k}(t)]^q \mu(B(x, t))^{q-1} t^{n-1} dt = L^n A_L^q \int W_k^\mu d\mu.$$

Let $\nu = L^{n/q} A_L \mu$. Then ν is a measure on E and

$$\int W_k^\nu d\nu \leq \int W_k^{\mu'} d\mu' \leq A.$$

Thus by Corollary 5.1.1 and (5.3.2)

$$C_{k,p}(E) \geq A \|\nu\|^p = A[L^{n/q} A_L]^p \|\mu\|^p = A L^{n(p-1)} A_L^p C_{k,p}(E').$$

This implies the required inequality. The theorem is proved. $\square$

6. Capacity strong type inequality

6.1. Weak maximum principle. Let $k(t) \not\equiv 0$ be a nonnegative nonincreasing l.s.c. function on $t > 0$ such that $\lim_{t\to\infty} k(t) = 0$. With a slight abuse of notation we write $k(x) = k(|x|)$ for $x \in \mathbb{R}^n$. We assume that k is locally integrable, i.e.

$$\int_0^\infty k(t) t^{n-1} dt < \infty.$$

Let $1 < p < \infty$. For a measure μ we define the *nonlinear potential*

$$u(x) = u_\mu(x) = k * (k * \mu)^{1/(p-1)}(x).$$

This potential satisfies the *weak maximum principle* below.

For a unit vector e we let

$$\Gamma_e = \{x : x \cdot e > |x| \cos \frac{\pi}{6}\}.$$

This is the cone with vertex at 0, axis along e and opening $\pi/6$. By inspection we find Q unit vectors $e_1, \ldots, e_Q$ such that $\mathbb{R}^n \setminus \{0\} = \Gamma_{e_1} \cup \cdots \cup \Gamma_{e_Q}$, where Q is an integer depending only on the dimension n.

THEOREM 6.1.1. *([4, Theorem 2.3]) There is a positive constant M depending only on the dimension n and p such that*

$$u(x) \leq M \sup_{y \in \text{supp} \mu} u(y).$$

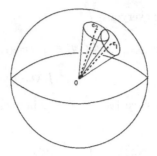

FIGURE 6.1

In fact, the constant M can be taken as

$$M = \begin{cases} Q^{1/(p-1)} & \text{if } 1 < p \le 2, \\ \\ Q & \text{if } 2 \le p < \infty. \end{cases}$$

In the proof we require several lemmas.

LEMMA 6.1.1. *Let $x \in \Gamma_e$ and let $P = \{y : (y - \frac{1}{2}x) \cdot x = 0\}$ be the perpendicular bisector of the segment joining 0 and x. Then P divides $\mathbb{R}^n$ into half spaces $P_0 = \{y : (y - \frac{1}{2}x) \cdot x \le 0\}$ and $P_1 = \{y : (y - \frac{1}{2}x) \cdot x \ge 0\}$. If $y \in \Gamma_e \cap P_0$, then $|y| < |x|$.*

PROOF. Let $y \in \Gamma_e \cap P_0$. Let θ be the angle between the segments $0x$ and $0y$ at 0. By definition $0 \le \theta < \pi/3$ and hence

$$\frac{1}{2}|x|^2 \ge x \cdot y = |x||y| \cos\theta > \frac{1}{2}|x||y|.$$

The lemma follows. $\square$

LEMMA 6.1.2. *Let P be an $(n-1)$-dimensional hyperplane in $\mathbb{R}^n$ and let P_+, P_- be the respective closed half-spaces bounded by P. The symbol x_+ designates a point in P_+ and x_- the symmetric point in P_-. If f is a nonnegative measurable function on $\mathbb{R}^n$ such that*

(6.1.1) $f(x_-) \le f(x_+)$ *for all $x_+ \in P_+$,*

then

(6.1.2) $k * f(x_-) \le k * f(x_+)$ *for all $x_+ \in P_+$.*

PROOF. In view of the monotone convergence theorem, we may assume that k and f are bounded and have compact support. Suppose $x_+ \in P_+$. Then $|x_+ - y_+| \le |x_- - y_+|$ and so $k(x_+ - y_+) - k(x_- - y_+) \ge 0$ for $y_+ \in P_+$. Hence, by (6.1.1)

$$\int_{P_+} [k(x_+ - y_+) - k(x_- - y_+)]f(y_+)dy_+ \ge \int_{P_+} [k(x_+ - y_+) - k(x_- - y_+)]f(y_-)dy_+$$

$$= \int_{P_-} [k(x_- - y_-) - k(x_+ - y_-)]f(y_-)dy_-,$$

where the equality follows since $|x_+ - y_+| = |x_- - y_-|$ and $|x_- - y_+| = |x_+ - y_-|$. Thus (6.1.2) follows. $\square$

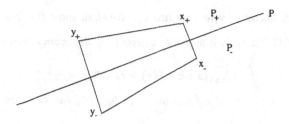

FIGURE 6.2

LEMMA 6.1.3. *Let P, P_+, P_-, x_+ and x_- be as in Lemma 6.1.2. If $\mu \geq 0$ is concentrated in P_+, then*

$$u_\mu(x_-) \leq u_\mu(x_+) \quad \text{for all } x_+ \in P_+.$$

PROOF. Let $x_+ \in P_+$. Then $|x_+ - y| \leq |x_- - y|$ for all $y \in P_+$. Hence

$$k * \mu(x_+) = \int_{P_+} k(x_+ - y)d\mu(y) \geq \int_{P_+} k(x_- - y)d\mu(y) = k * \mu(x_-).$$

Thus letting $f = (k * \mu)^{1/(p-1)}$ we have (6.1.1). Hence the required inequality follows from Lemma 6.1.2 $\square$

PROOF OF THEOREM 6.1.1. We may assume that $x \in \mathbb{R}^n \setminus \text{supp}\,\mu$. Let $\Gamma_{e_1}, \ldots, \Gamma_{e_Q}$ be the cones before Theorem 6.1.1. We let $\Gamma_j = x + \Gamma_{e_j}$. This is a cone with vertex at x and $\Gamma_1 \cup \cdots \cup \Gamma_Q = \mathbb{R}^n \setminus \{x\}$. Let $\mu_j = \mu|_{\Gamma_j}$ and $u_j = k * (k * \mu_j)^{1/(p-1)}$. Then

$$u(x) \leq \begin{cases} Q^{(p-2)/(p-1)} \sum_{j=1}^Q u_j(x) & \text{if } 1 < p \leq 2, \\ \\ \sum_{j=1}^Q u_j(x) & \text{if } 2 \leq p < \infty \end{cases}$$

(see the remark below). If we can show that

(6.1.3) $$u_j(x) \leq \sup_{y \in \text{supp}\,\mu} u(y),$$

then the theorem will readily follow. Let us prove (6.1.3). We may assume $\mu_j \neq 0$. Let $x_j \in \text{supp}\,\mu_j$ with $|x_j - x| = \text{dist}(x, \text{supp}\,\mu_j)$. If P is the perpendicular bisector of the segment joining x to x_j, then by Lemma 6.1.1 $\text{supp}\,\mu_j$ lies in the half space containing x_j. Hence Lemma 6.1.3 yields $u_j(x) \leq u_j(x_j)$, so that (6.1.3) follows. The theorem is proved. $\square$

Remark. Let $a_j \geq 0$. Then

$$(a_1 + \cdots + a_Q)^r \leq \begin{cases} a_1^r + \cdots + a_Q^r & \text{if } 0 < r \leq 1, \\ \\ Q^{r-1}(a_1^r + \cdots + a_Q^r) & \text{if } 1 \leq r < \infty. \end{cases}$$

The case when $0 < r \leq 1$ is proved by induction; the case when $1 \leq r < \infty$ is proved by Jensen's inequality. Apply this observation to $a_j = k * \mu_j$ and $r = 1/(p-1)$.

6.2. Capacity strong type inequality. We shall show the following theorem.

THEOREM 6.2.1. *There is a positive constant A_p depending only on n and p such that*

$$\int_0^\infty C_{k,p}(\{x : k * f(x) \geq t\})dt^p \leq A_p \|f\|_p^p$$

for any nonnegative measurable function f. In fact, A_p can be taken as

$$A_p = \begin{cases} (\dfrac{p}{p-1})^p M^{p-1} = (\dfrac{p}{p-1})^p Q & \text{if } 1 < p \leq 2, \\[4mm] p(\dfrac{p}{p-1})^{p-1} M = p(\dfrac{p}{p-1})^{p-1} Q & \text{if } 2 \leq p < \infty, \end{cases}$$

where M and Q are as in Theorem 6.1.1.

This theorem was established first by Hansson [33]. Later Maz'ya [52, Theorem 8.2.3] and Adams [2, Theorem 1.6] gave a simplified proof. Maz'ya and Adams used the joint measurability of $k * \mu_t(x)$ in (x, t), where μ_t is the capacitary measure for the set $\{x : k * f(x) \geq t\}$. However, the measurability does not seem to be obvious. We shall give an elementary proof which gets around this difficulty. We remark that Adams and Hedberg [3, Theorem 7.1.1] gave a beautiful short proof.

As an easy corollary to Theorem 6.2.1, we obtain the following characterization of Carleson type measures.

COROLLARY. *For a nonnegative measure μ the following statements are equivalent:*
(i) *For any $f \geq 0$ in $L^p(\mathbb{R}^n)$ we have $\int (k * f)^p d\mu \leq A \int f^p dx$.*
(ii) *For any compact set $K \subset \mathbb{R}^n$ we have $\mu(K) \leq BC_{k,p}(K)$.*

PROOF. (i) $\Longrightarrow$ (ii): This part can be proved without the capacity strong type inequality. Let K be a compact set. Take $f \geq 0$ such that $k * f \geq 1$ on K. Then

$$\mu(K) \leq \int (k * f)^p d\mu \leq A \int f^p dx.$$

Taking the infimum with respect to f, we obtain $\mu(K) \leq AC_{k,p}(K)$.

(ii) $\Longrightarrow$ (i): By the capacitability we have $\mu(E) \leq BC_{k,p}(E)$ for every Borel set E. Let $f \geq 0$ and apply the above inequality to $E = \{x : k * f(x) \geq t\}$. By Theorem 6.2.1 we have

$$\int f^p dx \geq \frac{1}{A_p} \int_0^\infty C_{k,p}(\{x : k * f(x) \geq t\})dt^p$$

$$\geq \frac{1}{A_p B} \int_0^\infty \mu(\{x : k * f(x) \geq t\})dt^p = \frac{1}{A_p B} \int (k * f)^p d\mu.$$

$\square$

6.3. Lemmas. In this section we state some elementary lemmas. Throughout this section we let $1 \leq q < \infty$.

LEMMA 6.3.1. *If $0 \leq S \leq T$, then*

$$q(T - S)S^{q-1} \leq T^q - S^q \leq q(T - S)T^{q-1}.$$

PROOF. Apply the mean value theorem to the function $f(t) = t^q$. $\square$

LEMMA 6.3.2. *If $a_j \geq 0$ for $-\infty < j < \infty$, then*

$$(6.3.1) \qquad q \sum_{i=-\infty}^{\infty} a_i (\sum_{j=i+1}^{\infty} a_j)^{q-1} \leq (\sum_{j=-\infty}^{\infty} a_j)^q \leq q \sum_{i=-\infty}^{\infty} a_i (\sum_{j=i}^{\infty} a_j)^{q-1},$$

$$(6.3.2) \qquad q \sum_{i=-\infty}^{\infty} a_i (\sum_{j=-\infty}^{i-1} a_j)^{q-1} \leq (\sum_{j=-\infty}^{\infty} a_j)^q \leq q \sum_{i=-\infty}^{\infty} a_i (\sum_{j=-\infty}^{i} a_j)^{q-1}.$$

PROOF. We prove (6.3.1) only; (6.3.2) can be proved in a similar way. Let N be an integer and let $S_i = \sum_{j=i}^{N} a_j$ for $i \leq N$ and $S_{N+1} = 0$. Apply Lemma 6.3.1 to $T = S_i$ and $S = S_{i+1}$ for $i \leq N$ to obtain

$$q a_i (\sum_{j=i+1}^{N} a_j)^{q-1} \leq S_i^q - S_{i+1}^q \leq q a_i (\sum_{j=i}^{N} a_j)^{q-1}.$$

Adding the above inequalities for $i = M, \ldots, N$, $(M < N)$, we obtain

$$q \sum_{i=M}^{N} a_i (\sum_{j=i+1}^{N} a_j)^{q-1} \leq S_M^q = (\sum_{j=M}^{N} a_j)^q \leq q \sum_{i=M}^{N} a_i (\sum_{j=i}^{N} a_j)^{q-1}.$$

Letting $N \to \infty$ and $M \to -\infty$, we obtain (6.3.1) by the monotone convergence theorem. □

LEMMA 6.3.3. *Let $1 < q \leq 2$ and let $a_j = a_j(x)$ be a nonnegative measurable function for $-\infty < j < \infty$. Then*

$$\int (\sum_{j=-\infty}^{\infty} a_j)^q dx \leq q (\int \sum_{i=-\infty}^{\infty} a_i^q dx)^{2-q} (\int \sum_{i=-\infty}^{\infty} a_i^{q-1} (\sum_{j=i}^{\infty} a_j) dx)^{q-1}.$$

PROOF. Write the right hand side of (6.3.1) as

$$q \sum_{i=-\infty}^{\infty} a_i^{q(2-q)} \cdot a_i^{(q-1)^2} (\sum_{j=i}^{\infty} a_j)^{q-1}$$

and apply Hölder's inequality with $1/(2-q) > 1$ and its conjugate $1/(q-1)$. □

LEMMA 6.3.4. *Let $2 \leq q < \infty$ and let $a_j = a_j(x)$ be a nonnegative measurable function for $-\infty < j < \infty$. Then*

$$\int (\sum_{j=-\infty}^{\infty} a_j)^q dx \leq q \sum_{i=-\infty}^{\infty} (\sum_{j=-\infty}^{i} \{\int a_i a_j^{q-1} dx\}^{1/(q-1)})^{q-1}.$$

PROOF. Observe $q - 1 \geq 1$. Apply Minkowski's inequality to the integral of the right hand side of (6.3.2) to obtain

$$q \int \sum_{i=-\infty}^{\infty} a_i (\sum_{j=-\infty}^{i} a_j)^{q-1} dx = q \sum_{i=-\infty}^{\infty} [\{\int a_i (\sum_{j=-\infty}^{i} a_j)^{q-1} dx\}^{1/(q-1)}]^{q-1}$$

$$\leq q \sum_{i=-\infty}^{\infty} [\sum_{j=-\infty}^{i} (\int a_i a_j^{q-1} dx)^{1/(q-1)}]^{q-1}.$$

□

6.4. Proof of Theorem 6.2.1. Hereafter we let $1 < p < \infty$ and $\dfrac{1}{p} + \dfrac{1}{q} = 1$. We observe that $1/(p-1) = q - 1$. For simplicity we write $k * (k * \mu)^{q-1}$ for the nonlinear potential of μ. It follows from Fubini's theorem that

$$\int k * (k * \mu)^{q-1} d\nu = \int (k * \mu)^{q-1} k * \nu dx.$$

In the sequel we shall use this identity frequently.

LEMMA 6.4.1. *Let* $2 \le p < \infty$ *and* $\alpha > 1$. *Suppose* μ_j *are measures such that* $k * (k * \mu_j)^{q-1} \le 1$ *on* $\operatorname{supp} \mu_j$ *for* $-\infty < j < \infty$. *Then*

$$\int \Big(\sum_{j=-\infty}^{\infty} \alpha^{(p-1)j} k * \mu_j \Big)^q dx \le q \Big(\frac{\alpha}{\alpha - 1} \Big)^{q-1} M^{q-1} \sum_{j=-\infty}^{\infty} \alpha^{pj} \|\mu_j\|,$$

where M *is the constant in Theorem 6.1.1.*

PROOF. In view of Theorem 6.1.1, we have $k * (k * \mu_j)^{q-1} \le M$ on $\mathbb{R}^n$. Apply Lemma 6.3.3 to $a_j = \alpha^{(p-1)j} k * \mu_j$ to obtain

$$\int \Big(\sum_{j=-\infty}^{\infty} \alpha^{(p-1)j} k * \mu_j \Big)^q dx$$

$$\le q \Big(\int \sum_{i=-\infty}^{\infty} (\alpha^{(p-1)i} k * \mu_i)^q dx \Big)^{2-q} \Big(\int \sum_{i=-\infty}^{\infty} (\alpha^{(p-1)i} k * \mu_i)^{q-1} \sum_{j=i}^{\infty} \alpha^{(p-1)j} k * \mu_j dx \Big)^{q-1}$$

$$= q \Big(\sum_{i=-\infty}^{\infty} \alpha^{pi} \int k * (k * \mu_i)^{q-1} d\mu_i \Big)^{2-q} \Big(\sum_{j=-\infty}^{\infty} \alpha^{(p-1)j} \sum_{i=-\infty}^{j} \alpha^i \int k * (k * \mu_i)^{q-1} d\mu_j \Big)^{q-1}$$

$$\le q \Big(\sum_{i=-\infty}^{\infty} \alpha^{pi} \|\mu_i\| \Big)^{2-q} \Big(\sum_{j=-\infty}^{\infty} \alpha^{(p-1)j} \sum_{i=-\infty}^{j} \alpha^i M \|\mu_j\| \Big)^{q-1}$$

$$= q M^{q-1} \Big(\sum_{i=-\infty}^{\infty} \alpha^{pi} \|\mu_i\| \Big)^{2-q} \Big(\sum_{j=-\infty}^{\infty} \alpha^{(p-1)j} \|\mu_j\| \frac{\alpha^{j+1}}{\alpha - 1} \Big)^{q-1}$$

$$= q \Big(\frac{\alpha}{\alpha - 1} \Big)^{q-1} M^{q-1} \sum_{j=-\infty}^{\infty} \alpha^{pj} \|\mu_j\|,$$

where we have used $\sum_{i=-\infty}^{j} \alpha^i = \alpha^{j+1}/(\alpha - 1)$. The lemma follows. $\square$

LEMMA 6.4.2. *Let* $1 < p \le 2$ *and* $\alpha > 1$ *Suppose* μ_j *are measures such that* $k * (k * \mu_j)^{q-1} \le 1$ *on* $\operatorname{supp} \mu_j$ *for* $-\infty < j < \infty$. *Then*

$$\int \Big(\sum_{j=-\infty}^{\infty} \alpha^{(p-1)j} k * \mu_j \Big)^q dx \le q \Big(\frac{\alpha^p}{\alpha^p - \alpha} \Big)^{q-1} M \sum_{j=-\infty}^{\infty} \alpha^{pj} \|\mu_j\|,$$

where M *is the constant in Theorem 6.1.1.*

PROOF. Apply Lemma 6.3.4 to $a_j = \alpha^{(p-1)j}k * \mu_j$ to obtain

$$\int(\sum_{j=-\infty}^{\infty} \alpha^{(p-1)j}k * \mu_j)^q dx$$

$$\leq q \sum_{i=-\infty}^{\infty}(\sum_{j=-\infty}^{i}\{\int \alpha^{(p-1)i}k * \mu_i(\alpha^{(p-1)j}k * \mu_j)^{q-1}dx\}^{1/(q-1)})^{q-1}$$

$$= q \sum_{i=-\infty}^{\infty} \alpha^{(p-1)i}(\sum_{j=-\infty}^{i} \alpha^{(p-1)j}\{\int k * (k * \mu_j)^{q-1}d\mu_i\}^{1/(q-1)})^{q-1}$$

$$\leq qM \sum_{i=-\infty}^{\infty} \alpha^{(p-1)i}\|\mu_i\|(\sum_{j=-\infty}^{i} \alpha^{(p-1)j})^{q-1}$$

$$= q(\frac{\alpha^p}{\alpha^p - \alpha})^{q-1}M \sum_{i=-\infty}^{\infty} \alpha^{pi}\|\mu_i\|,$$

where we have used $\sum_{j=-\infty}^{i} \alpha^{(p-1)j} = \frac{\alpha^p}{\alpha^p - \alpha}\alpha^{(p-1)i}$. The lemma follows. $\square$

LEMMA 6.4.3. *Let f be a nonnegative continuous function of compact support and E a Borel set. Let $\alpha > 1$, $E_j = \{x \in E : k * f(x) \geq \alpha^j\}$ and let μ_j be the capacitary measure for E_j, i.e.*

$$\text{supp } \mu_j \subset \overline{E_j}$$

$$k * (k * \mu_j)^{q-1} \geq 1 \ C_{k,p}\text{-a.e. on } E_j,$$

$$k * (k * \mu_j)^{q-1} \leq 1 \text{ on supp } \mu_j,$$

$$\|\mu_j\| = C_{k,p}(E_j).$$

Then

$$(1 - \alpha^{-p}) \sum_{j=-\infty}^{\infty} \alpha^{pj}\|\mu_j\| \leq \int_0^{\infty} C_{k,p}(\{x \in E : k * f(x) \geq t\})dt^p$$

$$\leq (\alpha^p - 1)\|f\|_p \left\|\sum_{j=-\infty}^{\infty} \alpha^{(p-1)j}k * \mu_j\right\|_q.$$

PROOF. By definition

(6.4.1)

$$C_{k,p}(E_{j+1})(\alpha^{p(j+1)} - \alpha^{pj}) \leq \int_{\alpha^j}^{\alpha^{j+1}} C_{k,p}(\{x : k * f(x) \geq t\})dt^p \leq C_{k,p}(E_j)(\alpha^{p(j+1)} - \alpha^{pj}).$$

The left hand side is equal to

$$(1 - \alpha^{-p})\alpha^{p(j+1)}C_{k,p}(E_{j+1}) = (1 - \alpha^{-p})\alpha^{p(j+1)}\|\mu_{j+1}\|,$$

whence adding this, we obtain the first required inequality. The right hand side of (6.4.1) is equal to

$$(\alpha^p - 1)\alpha^{pj}\|\mu_j\| \leq (\alpha^p - 1)\alpha^{pj}\int \frac{k * f}{\alpha^j}d\mu_j = (\alpha^p - 1)\alpha^{(p-1)j}\int f(k * \mu_j)dx.$$

Adding this we obtain

$$\int_0^{\infty} C_{k,p}(\{x \in E : k * f(x) \geq t\})dt^p \leq (\alpha^p - 1)\int f(\sum_{j=-\infty}^{\infty} \alpha^{(p-1)j}k * \mu_j)dx.$$

Hence Hölder's inequality yields the required inequality. $\square$

PROOF OF THEOREM 6.2.1. In view of the monotone convergence theorem, it is sufficient to show that

$$\int_0^\infty C_{k,p}(\{x \in B(0,r) : k * f(x) \geq t\})dt^p \leq A_p \|f\|_p^p,$$

for $r > 0$ and a nonnegative continuous function f with compact support, where A_p is independent of r and f. For simplicity we write I for the left hand side. Since $k * f$ is bounded, it follows that $\{x : k * f(x) \geq t\} = \emptyset$ for large t, say $t > T$, so that $I \leq C_{k,p}(B(0,r))T^p < \infty$. Let $\alpha > 1$ and let μ_j be the capacitary measure for $\{x \in B(0,r) : k * f(x) \geq \alpha^j\}$. By Lemma 6.4.3 with $E = B(0,r)$ we have

$$(6.4.2) \qquad I \leq (\alpha^p - 1) \|f\|_p \left\| \sum_{j=-\infty}^{\infty} \alpha^{(p-1)j} k * \mu_j \right\|_q.$$

Let M be the constant in Theorem 6.1.1.

Suppose $2 \leq p < \infty$. Then Lemma 6.4.1 and Lemma 6.4.3 yield

$$\left\| \sum_{j=-\infty}^{\infty} \alpha^{(p-1)j} k * \mu_j \right\|_q \leq q^{1/q} \left(\frac{\alpha}{\alpha - 1}\right)^{1/p} M^{1/p} \left(\sum_{j=-\infty}^{\infty} \alpha^{pj} \|\mu_j\| \right)^{1/q}$$

$$\leq q^{1/q} \left(\frac{\alpha}{\alpha - 1}\right)^{1/p} M^{1/p} (1 - \alpha^{-p})^{-1/q} I^{1/q}.$$

Since $I < \infty$, it follows from (6.4.2) that

$$I \leq (\alpha^p - 1)^p q^{p/q} \left(\frac{\alpha}{\alpha - 1}\right) M (1 - \alpha^{-p})^{1-p} \|f\|_p^p = q^{p-1}(\alpha^p - 1)\alpha^{p(p-1)} \frac{\alpha}{\alpha - 1} M \|f\|_p^p.$$

Letting $\alpha \to 1$, we obtain $I \leq pq^{p-1}M \|f\|_p^p$. Thus the required inequality follows.

Suppose $1 < p \leq 2$. Then Lemma 6.4.2 and Lemma 6.4.3 yield

$$\left\| \sum_{j=-\infty}^{\infty} \alpha^{(p-1)j} k * \mu_j \right\|_q \leq q^{1/q} \left(\frac{\alpha^p}{\alpha^p - \alpha}\right)^{1/p} M^{1/q} \left(\sum_{j=-\infty}^{\infty} \alpha^{pj} \|\mu_j\| \right)^{1/q}$$

$$\leq q^{1/q} \left(\frac{\alpha^p}{\alpha^p - \alpha}\right)^{1/p} M^{1/q} (1 - \alpha^{-p})^{-1/q} I^{1/q}.$$

Since $I < \infty$, it follows from (6.4.2) that

$$I \leq (\alpha^p - 1)^p q^{p/q} \left(\frac{\alpha^p}{\alpha^p - \alpha}\right)(\alpha^p - 1)^{1-p} \alpha^{p(p-1)} M^{p-1} \|f\|_p^p = q^{p-1}(\alpha^p - 1)\left(\frac{\alpha^{p^2}}{\alpha^p - \alpha}\right) M^{p-1} \|f\|_p^p.$$

Letting $\alpha \to 1$, we obtain $I \leq q^{p-1}\frac{p}{p-1}M^{p-1} \|f\|_p^p = q^p M^{p-1} \|f\|_p^p$. Thus the required inequality follows. $\square$

7. Quasiadditivity of capacity

7.1. Introduction. One of the most important properties of a capacity C is the countable subadditivity, i.e.

$$C(E) \leq \sum C(E_j), \quad E = \bigcup E_j.$$

We shall prove that the reverse inequality,

$$C(E) \geq A \sum C(E_j), \quad E = \bigcup E_j$$

up to a multiplicative constant, holds in a certain situation. Such a property will be referred to as *quasiadditivity*. We shall observe that the L^p-capacity and the Green capacity are quasiadditive with respect to a certain Whitney decomposition.

Let us begin with the Riesz capacity. Let $0 < \alpha < n$ and $k_\alpha(x) = |x|^{\alpha-n}$ the Riesz kernel on $\mathbb{R}^n$. Define the Riesz capacity by

$$R_{\alpha,p}(E) = \begin{cases} \inf\{\|f\|_p^p : k_\alpha * f \geq 1 \text{ on } E, f \geq 0\} & \text{if } 1 < p < \infty, \\[2mm] \inf\{\|\mu\| : k_\alpha * \mu \geq 1 \text{ on } E, \mu \geq 0\} & \text{if } p = 1. \end{cases}$$

In view of [30] we see that $R_{\alpha,1}(E)$ is equal to the usual (outer) α-capacity $C_\alpha(E)$. Quasiadditivity for decompositions into spherical shells has been considered by Landkof [40, Lemma 5.5 in p.304] and Adams [1, Theorem 7.5]. They showed

$$R_{\alpha,p}(E) \approx \sum_j R_{\alpha,p}(E_j) \quad \text{with } E_j = B(0, 2^{j+1}) \setminus B(0, 2^j).$$

Let $\{Q_k\}$ be the *Whitney decomposition* (cf. [62, p.16]) of the complement of the origin. We observe that each shell $B(0, 2^{j+1}) \setminus B(0, 2^j)$ is covered by N-many Whitney cubes Q_k with N depending only on dimension. Hence the result of Adams and Landkof may be regarded as the quasiadditivity for the Whitney decomposition. In the case of Green energy, quasiadditivity for the Whitney decomposition of a half space is discussed in Essén [26]. We shall show that the Whitney decomposition associated with a certain closed set has quasiadditivity. For this purpose we introduce a notion which is related to the dimension.

DEFINITION. Let F be a closed set having no interior points. Put $\delta(x) = \text{dist}(x, F)$ and let m_β be the measure defined by

$$m_\beta(E) = \int_E \delta(x)^{-\beta} dx.$$

We associate the least number $d = d(F)$ for which

$$(7.1.1) \qquad m_\beta(B(x,r)) \leq A_\beta r^{n-\beta}$$

holds for all $x \in F$ and $r > 0$ with a positive constant A_β, whenever $0 < \beta < n - d$.

It is easy to see that if $0 < \beta < n - d$, then

$$(7.1.2) \qquad m_\beta(B(x,r)) \approx \begin{cases} \delta(x)^{-\beta} r^n & \text{if } \delta(x) \geq 2r, \\ r^{n-\beta} & \text{if } \delta(x) \leq 2r. \end{cases}$$

The constant $d(F)$ is related to the dimension of F. In fact, if L is an m-dimensional affine subspace in $\mathbb{R}^n$, then $d(L) = m$. We can easily see that if F is an m-dimensional compact Lipschitz manifold, then $d(F) = m$. By definition if $F_1 \subset F_2$, then $d(F_1) \leq d(F_2)$. The Hausdorff dimension of F is not greater than $d(F)$. They are, in general, different; if $F = \{0\} \cup \bigcup_{j=1}^\infty \{(j^{-1}, 0, \ldots, 0)\}$, then $d(F) > 0$ and yet the Hausdorff dimension of F is equal to 0.

Our main result is

THEOREM 7.1.1. *Let $1 \leq p < \infty$ and suppose $\alpha p + d(F) < n$. Let $\{Q_j\}$ be the Whitney decomposition of $\mathbb{R}^n \setminus F$. Then, for any set $E \subset \mathbb{R}^n$,*

$$R_{\alpha,p}(E) \geq A \sum_j R_{\alpha,p}(E_j)$$

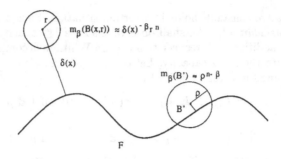

FIGURE 7.1

holds with $E_j = E \cap Q_j$ for some positive constant A.

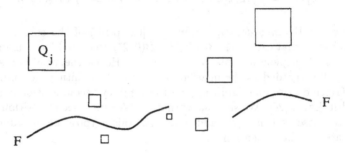

FIGURE 7.2

Let us note that $R_{\alpha,p}(F) = 0$ since the Hausdorff dimension of F is not greater than $d(F) < n - \alpha p$ (see [53, Theorem 21]). Since $d(\{0\}) = 0$, we see that Theorem 7.1.1 is a generalization of the aforementioned results of Landkof and Adams. Our proof is completely different; it relies on the following comparison between the Riesz capacity $R_{\alpha,p}$ and the measure $m_{\alpha p}$.

THEOREM 7.1.2. *Let $1 \le p < \infty$ and $\alpha p + d(F) < n$. If E is measurable, then*

$$m_{\alpha p}(E) \le A R_{\alpha,p}(E)$$

for some positive constant A.

In order to make our argument more clear, we shall consider a more general situation. Through such a generalization, we get a better understanding of the mechanism which yields Theorem 7.1.1 from Theorem 7.1.2. Let X be a locally compact Hausdorff space and let k be a nonnegative lower semicontinuous function on $X \times X$. Define the capacity C_k by

$$C_k(E) = \inf\{\|\mu\| : k(\cdot, \mu) \ge 1 \text{ on } E\},$$

where $k(\cdot, \mu) = \int_X k(\cdot, y) d\mu(y)$. Moreover, suppose λ is a fixed measure on X. For $p > 1$ we define

$$C_{k,p}(E) = \inf\{\|f\|_p^p : k(\cdot, f\lambda) \ge 1 \text{ on } E\}.$$

Obviously, if $X = \mathbb{R}^n$, $k(x, y) = k_\alpha(x - y)$ and λ is the Lebesgue measure, then $C_{k,p}(E) = R_{\alpha,p}(E)$. With a slight abuse of notation, we write $C_{k,1}(E)$ for $C_k(E)$.

DEFINITION. Let $\{Q_j\}$ and $\{Q_j^*\}$ be families of Borel subsets of X such that:

  (i) $Q_j \subset Q_j^*$,
  (ii) $X = \bigcup Q_j$,
  (iii) Q_j^* do not overlap so often, i.e., $\sum \chi_{Q_j^*} \leq N$.

Then we say that $\{Q_j, Q_j^*\}$ (or more simply $\{Q_j\}$) is a *quasidisjoint decomposition* of X. (We note that we do not exclude the possibility that sets from $\{Q_j\}$ are not mutually disjoint.)

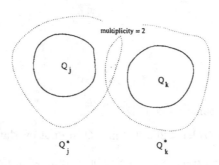

FIGURE 7.3

DEFINITION. Let σ be a (Borel) measure on X. We say that the measure σ is *comparable to* $C_{k,p}$ *with respect to* $\{Q_j\}$ if

$$\sigma(Q_j) \approx C_{k,p}(Q_j) \text{ for every } Q_j,$$
$$\sigma(E) \leq AC_{k,p}(E) \text{ for every Borel set } E.$$

DEFINITION. We say that the kernel k has the *Harnack property* with respect to the decomposition $\{Q_j, Q_j^*\}$ if

$$k(x,y) \approx k(x',y)$$

for $x, x' \in Q_j$ and $y \in X \setminus Q_j^*$.

The following is a generalization of Theorem 7.1.1.

THEOREM 7.1.3. *Let* $\{Q_j, Q_j^*\}$ *be a quasidisjoint decomposition of* X. *Suppose* k *has the Harnack property with respect to* $\{Q_j, Q_j^*\}$. *If there is a measure* σ *comparable to* $C_{k,p}$ *with respect to* $\{Q_j, Q_j^*\}$, *then for every* $E \subset X$

$$C_{k,p}(E) \approx \sum C_{k,p}(E \cap Q_j).$$

PROOF. For simplicity we shall consider only the case $p = 1$. The case when $p > 1$ can be proved similarly. The proof is rather easy. It is sufficient to show that C_k inherits the additivity of σ. Since we know that $C_k(E) \leq \sum C_k(E \cap Q_j)$, we have only to show the reverse inequality up to a multiplicative constant. Hence we may assume that $C_k(E) < \infty$. By definition we can find a measure μ such that $k(\cdot, \mu) \geq 1$ on E and $\|\mu\| \leq 2C_k(E)$. For each Q_j we decompose the measure μ into

$$\mu_j = \mu|_{Q_j^*}, \qquad \mu_j' = \mu|_{X \setminus Q_j^*}.$$

We have the following two cases:

  (a) $k(x, \mu_j) \geq \dfrac{1}{2}$ for all $x \in E \cap Q_j$.

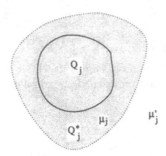

FIGURE 7.4

(b) $k(x, \mu_j') \geq \dfrac{1}{2}$ for some $x \in E \cap Q_j$.

If (a) holds, then $C_k(E \cap Q_j) \leq 2\|\mu_j\|$. Since Q_j^* do not overlap so often, we obtain

$$\sum{}' C_k(E \cap Q_j) \leq 2\sum{}' \|\mu_j\| \leq A\|\mu\| \leq AC_k(E),$$

where $\sum'$ denotes the summation over all Q_j for which (a) holds. In the second inequality we have used the fact that Q_j^* do not overlap so often. If (b) holds, then the Harnack property of k yields that

$$k(\cdot, \mu) \geq k(\cdot, \mu_j') \geq k(x, \mu_j') \geq A \text{ on } Q_j,$$

so that

$$k(\cdot, \mu) \geq A \text{ on } \bigcup{}'' Q_j,$$

where $\bigcup''$ denotes the union over all Q_j for which (b) holds. Hence

$$C_k(\bigcup{}'' Q_j) \leq A\|\mu\| \leq 2AC_k(E).$$

Since σ is comparable to C_k, it follows from the countable additivity of σ that

$$\sum{}'' C_k(E \cap Q_j) \leq \sum{}'' C_k(Q_j) \leq A\sum{}'' \sigma(Q_j)$$
$$\leq A\sigma(\bigcup{}'' Q_j) \leq AC_k(\bigcup{}'' Q_j) \leq AC_k(E).$$

This completes the proof. $\square$

COROLLARY 7.1.1. *Let Q_j, k and σ be as in Theorem 7.1.3. If E is a union of Q_j, then $C_{k,p}(E) \approx \sigma(E)$.*

PROOF OF THEOREM 7.1.1. Observe that the Whitney decomposition $\{Q_j\}$ and its expansion $\{Q_j^*\}$ gives a quasidisjoint decomposition of $\mathbb{R}^n \setminus F$ ([62, p.169]). The Riesz kernel k_α enjoys the Harnack property. By (7.1.2), $m_{\alpha p}(Q_j) \approx R_{\alpha,p}(Q_j)$ and hence Theorem 7.1.2 says that $m_{\alpha p}$ is comparable to $R_{\alpha,p}$. We know that $R_{\alpha,p}(F) = 0$. Thus Theorem 7.1.1 follows from Theorem 7.1.2 and Theorem 7.1.3. $\square$

COROLLARY 7.1.2. *Let $p \geq 1$ and $\alpha p + d(F) < n$. If E is a union of Whitney cubes, then $R_{\alpha,p}(E) \approx m_{\alpha p}(E)$.*

7.2. How do we get a comparable measure? In the proof of Theorem 7.1.3, a comparable measure has played an important role. However, how do we obtain a comparable measure? This is the most difficult and the most interesting question. It seems there is no unified argument.

Our Theorem 7.1.2 follows from the following norm estimate.

THEOREM 7.2.1. *Let $p \geq 1$ and $\alpha p + d(F) < n$.*
(i) If $p > 1$, then

$$\int_{\mathbb{R}^n} |k_\alpha * f|^p dm_{\alpha p} \leq A \|f\|_p^p.$$

(ii) If $p = 1$, then for $\lambda > 0$

$$m_\alpha(\{x \in \mathbb{R}^n : k_\alpha * \mu(x) > \lambda\}) \leq A \frac{\|\mu\|}{\lambda}.$$

PROOF OF THEOREM 7.1.2. Suppose $p > 1$. Take $f \geq 0$ such that $k_\alpha * f \geq 1$ on E. Then by Theorem 7.2.1 (i)

$$m_{\alpha p}(E) \leq \int_{k_\alpha * f \geq 1} (k_\alpha * f)^p dm_{\alpha p} \leq A \|f\|_p^p.$$

Taking the infimum with respect to such f, we obtain $m_{\alpha p}(E) \leq A R_{\alpha, p}(E)$. The case when $p = 1$ similarly follows from Theorem 7.2.1 (ii). $\square$

Remark. Actually, $m_{\alpha p}(E) \leq A R_{\alpha, p}(E)$ for every Borel set E and the norm estimate in Theorem 7.2.1 (i) are equivalent for $p > 1$. This follows from the *capacity strong type inequality*: for $f \geq 0$

$$\int_0^\infty R_{\alpha, p}(\{k_\alpha * f \geq t\}) dt^p \leq A \|f\|_p^p$$

(see [33], [52, Theorem 8.2.3] and [2, Theorem 1.6]). In fact,

$$\int |k_\alpha * f|^p dm_{\alpha p} \leq \int_0^\infty m_{\alpha p}(\{k_\alpha * |f| \geq t\}) dt^p \leq A \int_0^\infty R_{\alpha, p}(\{k_\alpha * |f| \geq t\}) dt^p \leq A \|f\|_p^p.$$

In [10], Theorem 7.2.1 was proved by using a certain good λ inequality. Here we give a different proof based on the duality. This proof works only for $p > 1$ but is simpler. I would like to acknowledge that Alexander Volberg gave me the idea of the proof.

We need several definitions. For $0 < \alpha < n$ we let

$$M_\alpha f(x) = \sup_{x \in Q} \frac{1}{|Q|^{(n-\alpha)/n}} \int_Q |f| dy,$$

where the supremum is taken over all cubes containing x. This is the fractional maximal function. We define

$$\mathcal{M}_{\alpha p} f(x) = \sup_{x \in Q} \frac{1}{m_{\alpha p}(Q)} \int_Q |f| dm_{\alpha p},$$

where the supremum is taken over all cubes containing x. This is the maximal function with respect to the measure $m_{\alpha p}$. Moreover let

$$f^\#(x) = \sup_{x \in Q} \frac{1}{|Q|} \int_Q |f(y) - f_Q| dy,$$

where the supremum is taken over all cubes containing x, and $f_Q = |Q|^{-1} \int_Q f dy$. This is the Fefferman-Stein #-maximal function. Finally we write in general

$$\|f\|_{q,\beta} = (\int |f|^q dm_\beta)^{1/q}.$$

PROOF OF THEOREM 7.2.1 (I). Let $1/p + 1/q = 1$. By duality

$$\|k_\alpha * f\|_{p,\alpha p} = \sup_{\|g\|_{q,\alpha p} \le 1} |\int (k_\alpha * f) g dm_{\alpha p}|.$$

By Fubini's theorem and Hölder's inequality

$$|\int (k_\alpha * f) g dm_{\alpha p}| = |\int k_\alpha * (g \delta^{-\alpha p}) f dy| \le \|k_\alpha * (g \delta^{-\alpha p})\|_q \|f\|_p.$$

Therefore, if we have

(7.2.1) $$\|k_\alpha * (g \delta^{-\alpha p})\|_q \le A \|g\|_{q,\alpha p},$$

then

$$\|k_\alpha * f\|_{p,\alpha p} \le \sup_{\|g\|_{q,\alpha p} \le 1} A \|g\|_{q,\alpha p} \|f\|_p = A \|f\|_p.$$

Thus the theorem follows.

Now let us prove (7.2.1). In view of the norm inequality $\|F^{\#}\|_q \approx \|F\|_q$ (cf [29]), it is sufficient to show that

(7.2.2) $$\|(k_\alpha * (g \delta^{-\alpha p}))^{\#}\|_q \le A \|g\|_{q,\alpha p}.$$

We have the pointwise estimate

(7.2.3) $$|(k_\alpha * (g \delta^{-\alpha p}))^{\#}(x)| \le A M_\alpha (g \delta^{-\alpha p})(x)$$

(see Exercise 7.2.1 below or [2, Theorem 2.2]). By definition

(7.2.4) $$M_\alpha (g \delta^{-\alpha p})(x) \le M_\alpha (\delta^{-\alpha p})(x) \mathcal{M}_{\alpha p} g(x).$$

In fact, for a cube Q containing x,

$$\frac{1}{|Q|^{(n-\alpha)/n}} \int_Q |g \delta^{-\alpha p}| dy = \frac{m_{\alpha p}(Q)}{|Q|^{(n-\alpha)/n}} \frac{1}{m_{\alpha p}(Q)} \int_Q |g| dm_{\alpha p} \le M_\alpha (\delta^{-\alpha p})(x) \mathcal{M}_{\alpha p} g(x).$$

By the estimate (7.1.2) we have

(7.2.5) $$M_\alpha (\delta^{-\alpha p})(x) \le A \delta(x)^{\alpha(1-p)}.$$

Collecting (7.2.3), (7.2.4) and (7.2.5) altogether, we obtain

$$|(k_\alpha * (g \delta^{-\alpha p}))^{\#}(x)|^q \le A \delta(x)^{-\alpha p} (\mathcal{M}_{\alpha p} g(x))^q,$$

so that

$$\|(k_\alpha * (g \delta^{-\alpha p}))^{\#}\|_q \le A \|\mathcal{M}_{\alpha p} g\|_{q,\alpha p} \le A \|g\|_{q,\alpha p},$$

where the last inequality is the strong (q,q) of the maximal function $\mathcal{M}_{\alpha p} g$. Thus (7.2.2) and hence (7.2.1) follows. The proof is complete. $\square$

Exercise 7.2.1. Let μ be a nonnegative measure. Prove $(k_\alpha\mu)^\#(x) \le AM_\alpha\mu(x)$. In fact the reverse inequality also holds ([2, Theorem 2.2]).

Hint. Let Q be a cube and Q^* its double. Let x be the center of Q and r the diameter of Q. Put $\mu' = \mu|_{Q^*}$ and $\mu'' = \mu - \mu'$. By Fubini's theorem

$$\int_Q k_\alpha * \mu' dx \le Ar^\alpha \mu(Q^*) \le Ar^n M_\alpha\mu(x),$$

so that

$$\frac{1}{|Q|}\int_Q |k_\alpha * \mu' - (k_\alpha * \mu')_Q| dy \le AM_\alpha\mu(x).$$

On the other hand, the mean value theorem yields

$$|k_\alpha(y-w) - k_\alpha(z-w)| \le Ar|x-w|^{\alpha-n-1}$$

for $y, z \in Q$ and $w \notin Q^*$. Hence if $y \in Q$, then

$$|k_\alpha * \mu''(y) - (k_\alpha * \mu'')_Q| = |\int_{\mathbb{R}^n\backslash Q^*}(k_\alpha(y-w) - \frac{1}{|Q|}\int_Q k_\alpha(z-w)dz)d\mu(w)|$$

$$\le Ar\int_{\mathbb{R}^n\backslash Q^*}|x-w|^{\alpha-n-1}d\mu(w) \le AM_\alpha\mu(x).$$

Remark. We do not know the above duality argument works for the case $p = 1$. The argument in [10] works for $p \ge 1$. There is another approach from two-weight norm inequalities (see Pérez [57]).

For completeness we state the weighted norm inequality in [10] and give a proof of Theorem 7.2.1 (ii). Let $T_\alpha f(x) = \int_{\mathbb{R}^n} k_\alpha(x-y)f(y)dm_\alpha(y)$.

THEOREM 7.2.2. *Let* $\alpha + d(F) < n$ *and* $1 < p < \infty$. *Then* $\|T_\alpha f\|_{p,\alpha} \le A\|f\|_{p,\alpha}$. *Moreover, if* w *satisfies the Muckenhoupt* A_p *condition with respect to* m_α, *then*

$$(7.2.6) \qquad \int_{\mathbb{R}^n}|T_\alpha f|^p w dm_\alpha \le A\int_{\mathbb{R}^n}|f|^p w dm_\alpha.$$

Observe that $\alpha p + d(F) < n$ if and only if $w = \delta^{\alpha(1-p)}$ satisfies the Muckenhoupt A_p condition with respect to m_α. Let $g = \delta^{-\alpha}f$ and let $w = \delta^{\alpha(1-p)}$. Then $T_\alpha f = k_\alpha * g$ and (7.2.6) becomes Theorem 7.2.1 (i). Theorem 7.2.1 (ii) can be deduced from Theorem 7.2.2 with $p = 2$. Since this process was only sketched in [10], we give a complete proof below.

PROOF OF THEOREM 7.2.1 (II). As was observed in [10], we may assume that μ is absolutely continuous and $d\mu = f dx$ with $f \ge 0$. By the *Calderón-Zygmund lemma* (e.g. [62, p.17]) we have a family of cubes Q_j whose interior Q_j° are mutually disjoint with the following properties:

(i) $f(x) \le \lambda$ a.e. on $\mathbb{R}^n \backslash \Omega$ with $\Omega = \cup_j Q_j$.
(ii) For each cube Q_j

$$\lambda < \frac{1}{m_\alpha(Q_j)}\int_{Q_j} f dm_\alpha \le A\lambda.$$

Let

$$g(x) = \begin{cases} f(x) & \text{on } \mathbb{R}^n \backslash \Omega, \\ \frac{1}{m_\alpha(Q_j)}\int_{Q_j} f dm_\alpha & \text{on } Q_j^\circ. \end{cases}$$

and $b = f - g$. Obviously, $\|g\|_{1,\alpha} \leq \|f\|_{1,\alpha}$, $\|b\|_{1,\alpha} \leq 2\|f\|_{1,\alpha}$ and $m_\alpha(\Omega) \leq \|f\|_{1,\alpha}/\lambda$. Since

$$m_\alpha(\{|T_\alpha f| > \lambda\}) \leq m_\alpha(\{|T_\alpha g| > \frac{\lambda}{2}\}) + m_\alpha(\{|T_\alpha b| > \frac{\lambda}{2}\}),$$

we estimate each term in the right hand side separately. First we observe

$$\|g\|_{2,\alpha}^2 = \int_{\mathbb{R}^n \setminus \Omega} |g|^2 dm_\alpha + \int_\Omega |g|^2 dm_\alpha \leq \lambda \int_{\mathbb{R}^n \setminus \Omega} f dm_\alpha + A\lambda^2 m_\alpha(\Omega) \leq (1+A)\lambda\|f\|_{1,\alpha}.$$

Hence by Theorem 7.2.2 with $p = 2$

$$(\frac{\lambda}{2})^2 m_\alpha(\{|T_\alpha g| > \frac{\lambda}{2}\}) \leq \int |T_\alpha g|^2 dm_\alpha \leq A \int |g|^2 dm_\alpha \leq A\lambda\|f\|_{1,\alpha}.$$

Thus, $m_\alpha(\{|T_\alpha g| > \lambda/2\}) \leq A\|f\|_{1,\alpha}/\lambda$. To treat the second term let y_j be the center of Q_j and $\tilde{Q}_j$ the double of Q_j. We put $\tilde{\Omega} = \cup_j \tilde{Q}_j$. By an elementary calculation we have

$$\sup_{y \in Q_j} \int_{x \notin \tilde{Q}_j} |k_\alpha(x-y) - k_\alpha(x-y_j)| dm_\alpha(x) \leq A,$$

where A is independent of Q_j. Hence, it follows from $\int_{Q_j} b\, dm_\alpha = 0$ that

$$\int_{\mathbb{R}^n \setminus \tilde{\Omega}} |T_\alpha b| dm_\alpha \leq \sum_j \int_{x \notin \tilde{Q}_j} \int_{y \in Q_j} |k_\alpha(x-y) - k_\alpha(x-y_j)||b(y)| dm_\alpha(y) dm_\alpha(x)$$

$$\leq A \sum_j \int_{y \in Q_j} |b(y)| dm_\alpha(y) \leq A\|f\|_{1,\alpha}.$$

Therefore $m_\alpha(\{|T_\alpha b| > \lambda/2\}) \leq A\|f\|_{1,\alpha}/\lambda$, since

$$m_\alpha(\tilde{\Omega}) \leq A \sum_j m_\alpha(Q_j) = A m_\alpha(\Omega) \leq A\|f\|_{1,\alpha}/\lambda.$$

Thus $m_\alpha(\{T_\alpha f > \lambda\}) \leq A\|f\|_{1,\alpha}/\lambda$ for $d\mu = f dx$. The proof is complete. $\square$

7.3. Green energy. For capacities represented as Green energy we can get a comparable measure as follows. Let D be a domain with Green function $G(x,y)$. We say that $D \subset \mathbb{R}^n$ is *uniformly Δ-regular* if there are constants $r_0 > 0$ and ε_1, $0 < \varepsilon_1 < 1$, such that, for all $x \in \partial D$ and all $0 < r < r_0$,

$$w = w_{x,r} \leq 1 - \varepsilon_1 \quad \text{on } B(x, r/2) \cap D,$$

where $w = w_{x,r}$ is the harmonic measure of $\partial B(x,r) \cap D$ in the region $B(x,r) \cap D$ ([16, Definition 2]). It is easy to see that a Lipschitz domain and an NTA domain (cf. [38]) are both uniformly Δ-regular.

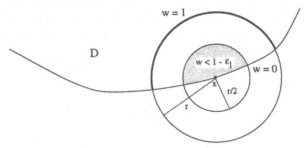

FIGURE 7.5

Let D be a uniformly Δ-regular domain and let $\{Q_j\}$ be the Whitney decomposition. Let u be a nonnegative superharmonic function on D. Suppose u has the Harnack property:

(7.3.1)
$$\sup_{Q_j} u \le A_0 \inf_{Q_j} u$$

with A_0 independent of Q_j. A typical example of u is given by $u = \min\{h^a, b\}$ with $0 < a \le 1$, $0 < b \le \infty$ and a positive harmonic function h.

For a compact subset K of D we let $\hat{R}_u^K$ be the regularized reduced function of u with respect to K. It is easy to see that $\hat{R}_u^K$ is a Green potential. Let $\hat{R}_u^K = v_K = G(\cdot, \lambda_u^K)$. The energy

$$\gamma_u(K) = \iint G(x,y)d\lambda_u^K(x)d\lambda_u^K(y)$$

is called the Green energy of K relative to u. For an open subset V we let

$$\gamma_u(V) = \sup\{\gamma_u(K) : K \text{ is compact}, K \subset V\},$$

and then for a general subset E

$$\gamma_u(E) = \inf\{\gamma_u(V) : V \text{ is open}, E \subset V\}.$$

The quantity $\gamma_u(E)$ is also called the Green energy relative to u. If $u \equiv 1$, then γ_u is the usual *Green capacity* C_G (see [40, pp.174–177]). If $D = \{x = (x_1, \ldots, x_n) : x_1 > 0\}$ and $u(x) = x_1$, then γ_u is the Green energy defined by Essén and Jackson [28, Definition 2.2]. Observe that the Green energy can be represented in different ways: as the Dirichlet integral and as the capacity treated in Theorem 7.1.3, i.e.

$$\gamma_u(K) = \int_D |\nabla v_K|^2 dx = C_k(K),$$

where $k(x,y) = G(x,y)/(u(x)u(y))$. Moreover, $\gamma_u(E) = C_k(E)$ for a general set E (see [30]). Thus our Theorem 7.1.3 can be applied to the Green energy. The kernel $k(x,y)$ is a generalization of Naïm's Θ kernel.

We define the measure σ_u on D by

$$\sigma_u(E) = \int_E \left(\frac{u(x)}{\delta(x)}\right)^2 dx.$$

We shall show that σ_u is comparable to γ_u with respect to $\{Q_j\}$. The following is easy from the assumption on u.

PROPOSITION 7.3.1. *By* $\operatorname{cap}(E)$ *we denote the logarithmic capacity of* E *if* $n = 2$, *and the Newtonian capacity of* E *if* $n \ge 3$. *Let* Q_j *be a Whitney cube and let* t_j, r_j *and* x_j *be the distance of* Q_j *from* ∂D, *the diameter of* Q_j *and the center of* Q_j, *respectively. If* $E \subset Q_j$, *then with* $u_j = u(x_j)$

$$\gamma_u(E) \approx \begin{cases} u_j^2 \left(\log \dfrac{4t_j}{\operatorname{cap}(E)}\right)^{-1} & \text{for } n = 2, \\ u_j^2 \operatorname{cap}(E) & \text{for } n \ge 3. \end{cases}$$

In particular,

$$\gamma_u(Q_j) \approx \sigma_u(Q_j) \approx u_j^2 r_j^{n-2}.$$

PROOF. For simplicity we consider only the case $n \ge 3$. Then $G(x,y) \approx |x - y|^{2-n}$ for $x, y \in Q_j$ and the estimate readily follows. $\square$

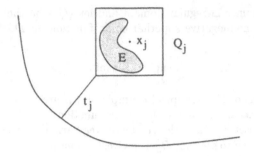

FIGURE 7.6

To compare C_k and σ for a general set, we need Hardy's inequality due to Ancona [16, (1)].

LEMMA 7.3.1. *If D is uniformly Δ-regular, then there is a positive constant A depending only on D such that*

$$\int_D \left|\frac{\psi(x)}{\delta(x)}\right|^2 dx \le A \int_D |\nabla\psi(x)|^2 dx$$

for all $\psi \in W_0^{1,2}(D)$.

Remark. Lewis [44] gave a different proof based on uniformly fat sets. His proof works for L^p-Hardy's inequality with $1 < p < \infty$. Wannebo [66] also gave a different proof.

PROPOSITION 7.3.2. *Let E be a Borel subset of D. Then*

$$\sigma_u(E) \le A\gamma_u(E).$$

PROOF. Let K be a compact subset of E and write $v_K = \hat{R}_u^K = G(\cdot, \lambda_u^K)$. Since $v_K = u$ q.e. on K, it follows from the Hardy inequality that

$$\gamma_u(E) \ge \gamma_u(K) = \int_D |\nabla v_K|^2 dx \ge A \int_D \left(\frac{v_K}{\delta}\right)^2 dx \ge A \int_K \left(\frac{u}{\delta}\right)^2 dx = A\sigma_u(K).$$

Hardy's inequality is used in the second inequality. Since $K \subset E$ is an arbitrary compact set, we have the required inequality. □

Remark. The above inequality is equivalent to Hardy's inequality. See for example Maz'ya' [52, §2.3].

THEOREM 7.3.1. *Suppose u satisfies (7.3.1). Then γ_u is quasiadditive with respect to $\{Q_j\}$, i.e.*

$$\gamma_u(E) \approx \sum \gamma_u(E \cap Q_j) \approx \begin{cases} \sum u_j^2 \left(\log \dfrac{4t_j}{\operatorname{cap}(E \cap Q_j)}\right)^{-1} & \text{for } n = 2, \\ \sum u_j^2 \operatorname{cap}(E \cap Q_j) & \text{for } n \ge 3. \end{cases}$$

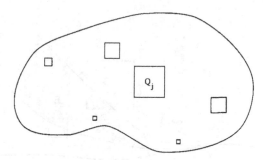

FIGURE 7.7

COROLLARY 7.3.1. *The Green capacity $C_G = \gamma_1$ is quasiadditive with respect to* $\{Q_j\}$, *i.e.*

$$C_G(E) \approx \sum C_G(E \cap Q_j) \approx \begin{cases} \sum \left(\log \dfrac{4t_j}{\operatorname{cap}(E \cap Q_j)} \right)^{-1} & \text{for } n = 2, \\ \sum \operatorname{cap}(E \cap Q_j) & \text{for } n \geq 3. \end{cases}$$

It seems this simple corollary is new even for a smooth domain.

COROLLARY 7.3.2. *Let $x_0 \in D$ and let $g(x) = \min\{G(x, x_0), 1\}$. Then γ_g is quasiadditive with respect to* $\{Q_j\}$, *i.e. with $g_j = g(x_j)$*

$$\gamma_g(E) \approx \sum \gamma_g(E \cap Q_j) \approx \begin{cases} \sum g_j^2 \left(\log \dfrac{4t_j}{\operatorname{cap}(E \cap Q_j)} \right)^{-1} & \text{for } n = 2, \\ \sum g_j^2 \operatorname{cap}(E \cap Q_j) & \text{for } n \geq 3. \end{cases}$$

7.4. Application. Here we give some applications to thin sets. We say that E is *α-thin* at y if there is a measure μ such that

$$\liminf_{\substack{x \to y \\ x \in E}} k_\alpha * \mu(x) > k_\alpha * \mu(y).$$

The following Wiener criterion is known: E is α-thin at y if and only if

$$\sum_{j=1}^{\infty} 2^{j(n-\alpha)} C_\alpha(B(y, 2^{1-j}) \setminus B(y, 2^{-j})) < \infty.$$

Moreover it is known that E is α-thin at y if and only if there is a potential $k_\alpha * \mu \not\equiv \infty$ such that

$$\lim_{\substack{x \to y \\ x \in E}} \frac{k_\alpha * \mu(x)}{k_\alpha(x - y)} = \infty.$$

(cf. [21, Theorem IX, 7] and [54, Theorem A]). From Theorem 7.1.1 with $p = 1$ we have the following *refined Wiener criterion.*

COROLLARY 7.4.1. *Let $\alpha + d(F) < n$ and let $y \in F$. Suppose E is a bounded set. Let $\{Q_j\}$ be the Whitney decomposition of $\mathbb{R}^n \setminus F$ and let $R_j(y) = \operatorname{dist}(y, Q_j)$. Then E is α-thin at y if and only if*

$$\sum_j \frac{C_\alpha(E \cap Q_j)}{R_j(y)^{n-\alpha}} < \infty.$$

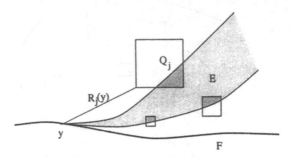

FIGURE 7.8

Moreover Theorem 7.1.2 says

$$\int_{E\cap Q_j} \delta(x)^{-\alpha} dx = m_\alpha(E\cap Q_j) \le A C_\alpha(E\cap Q_j).$$

Hence we have a necessary condition for α-thinness.

COROLLARY 7.4.2. *Let $\alpha + d(F) < n$ and let $y \in F$. If a bounded measurable set E is α-thin at y, then*

$$\int_E \frac{\delta(x)^{-\alpha}}{|x-y|^{n-\alpha}} dx < \infty.$$

Corollary 7.4.2 may be considered to be a counterpart of [24, Theorem 2]. Let us state a result corresponding to [24, Theorem 1]. A sequence $\{x_j\}$ is said to be separated if there is a positive constant ε such that

$$|x_j - x_k| \ge \varepsilon\delta(x_j) \quad \text{for } j \ne k.$$

It is easy to see that $\{x_j\}$ is separated if and only if the number of points x_j lying in Q_k is bounded by a positive constant independent of k.

COROLLARY 7.4.3. *Let $\alpha + d(F) < n$ and let $y \in F$. Suppose E is a bounded set. Let $\tilde{E} = \bigcup_{Q_j \cap E \ne \emptyset} Q_j$. Then the following statements are equivalent:*

(i) *$\tilde{E}$ is α-thin at y.*

(ii) $\displaystyle\int_{\tilde{E}} \frac{\delta(x)^{-\alpha}}{|x-y|^{n-\alpha}} dx < \infty.$

(iii) *E does not contain a separated sequence $\{x_j\}$ convergent to y such that*

$$\sum_j \left(\frac{\delta(x_j)}{|x_j - y|}\right)^{n-\alpha} = \infty.$$

(iv) $\displaystyle\sum_{Q_j \cap E \ne \emptyset} \left(\frac{r_j}{R_j(y)}\right)^{n-\alpha} < \infty$, *where r_j is the diameter of Q_j and $R_j(y) = \text{dist}(Q_j, y)$.*

(v) *There is a measure μ supported on F such that $k_\alpha * \mu \not\equiv \infty$ and*

$$\lim_{\substack{x \to y \\ x \in E}} \frac{k_\alpha * \mu(x)}{k_\alpha(x-y)} = \infty.$$

Let D be a $C^{1,\alpha}$-domain or more generally a Liapunov domain with Green function $G(x,y)$. Fix $x_0 \in D$ and let $g(x) = \min\{G(x,x_0),1\}$. In view of [67], we have

(7.4.1) $$g(x) \approx \delta(x) \approx \text{dist}(x,\partial D) \quad \text{for } x \in D.$$

Let $K(x,y) = G(x,y)/g(y)$. Then K extends continuously on $D \times \overline{D}$. The extension is denoted by the same symbol K and called the Martin kernel for D. For simplicity we write $K_y = K(\cdot,y)$. We say that $E \subset D$ is *minimally thin* at $y \in \partial D$, if $\hat{R}^E_{K_y} \neq K_y$. This is equivalent to

(7.4.2) $$\sum_{j=1}^{\infty} \hat{R}^{E_j}_{K_y}(x_0) < \infty,$$

where $E_j = E \cap B(y, 2^{1-j}) \setminus B(y, 2^{-j})$ (see [21, Chapter XVII]). By the estimate of Widman [67] we see that $K_y(x) \approx g(x)|x-y|^{-n}$. Hence

$$\hat{R}^{E_j}_{K_y}(x_0) \approx 2^{jn} \hat{R}^{E_j}_g(x_0).$$

Moreover, $\hat{R}^{E_j}_g(x_0) = \gamma_g(E_j)$, where γ_g is the Green energy relative to g. In fact, we may assume that E_j is a compact set K and write $\hat{R}^K_g = G\lambda^K_g$. it is well known that λ^K_g is supported on K and $G(x_0,\cdot) = g = \hat{R}^K_g$ q.e. on K. Hence

$$\hat{R}^K_g(x_0) = \int \hat{R}^K_g(y) d\lambda^K_g(y) = \iint G(x,y) d\lambda^K_g(x) d\lambda^K_g(y) = \gamma_g(K).$$

Thus (7.4.2) becomes

(7.4.3) $$\sum_{j=1}^{\infty} 2^{jn} \gamma_g(E_j) < \infty.$$

This Wiener type criterion for minimal thinness was essentially given by Lelong-Ferrand [42]. Our quasiadditivity (Corollary 7.3.2) gives a finer criterion

COROLLARY 7.4.4. *Let D be a $C^{1,\alpha}$-domain or more generally a Liapunov domain and $y \in \partial D$. Then $E \subset D$ is minimally thin at y if and only if*

$$\sum \frac{t_j^2}{R_j(y)^n} \text{cap}(E \cap Q_j) < \infty \qquad \text{if } n \geq 3,$$

$$\sum \frac{t_j^2}{R_j(y)^2} \left(\log \frac{4t_j}{\text{cap}(E \cap Q_j)}\right)^{-1} < \infty \qquad \text{if } n = 2,$$

where $t_j = \text{dist}(Q_j, \partial D)$ and $R_j(y) = \text{dist}(Q_j, y)$.

In the above Wiener type criterion the logarithmic capacity and the Newtonian capacity appear. Hayman asked whether it is possible to replace (7.4.3) by criteria using logarithmic capacity or Newtonian capacity. Our Corollary 7.4.4 gives an answer to his question.

Remark. It is possible to give a similar Wiener type criterion for minimal thinness for a Lipschitz domain and an NTA domain. The estimate of the Martin kernel is more delicate and the form of the criterion is rather complicated. For details see [12].

Remark. Slightly weaker forms of Corollary 7.3.2 and Corollary 7.4.4 were first given by Essén [26]. In [27, Section 1], he has given a proof of Corollary 7.4.4 based on the weak L^1-estimates of Sjögren [59].

In view of (7.4.1), Proposition 7.3.2 and Corollary 7.1.1 we have the following.

COROLLARY 7.4.5. *Let D be a $C^{1,\alpha}$-domain or more generally a Liapunov domain. Then $|E| \leq A\gamma_g(E)$. Moreover, if E is a union of Whitney cubes, then $|E| \approx \gamma_g(E)$.*

This corollary and (7.4.3) yield the following.

COROLLARY 7.4.6. *Let D be a $C^{1,\alpha}$-domain or more generally a Liapunov domain and $y \in \partial D$. If $E \subset D$ is minimally thin at y, then*

$$(7.4.4) \qquad \int_E \frac{dx}{|x-y|^n} dx < \infty.$$

Moreover, if E is a union of Whitney cubes, then (7.4.4) is also sufficient for E to be minimally thin at y.

The above corollary was given by Dahlberg [24]. See also [50], [51] and [59]. The following is the counterpart of Corollary 7.4.3 for minimal thinness.

COROLLARY 7.4.7. *Let D be a $C^{1,\alpha}$-domain or more generally a Liapunov domain and $y \in \partial D$. Let $E \subset D$ and $\tilde{E} = \bigcup_{Q_j \cap E \neq \emptyset} Q_j$. Then the following statements are equivalent:*

(i) *$\tilde{E}$ is minimally thin at y.*

(ii) $\displaystyle \int_{\tilde{E}} \frac{dx}{|x-y|^n} < \infty.$

(iii) *E does not contain a separated sequence $\{x_j\}$ convergent to y such that*

$$\sum_j \left(\frac{\delta(x_j)}{|x_j - y|}\right)^n = \infty.$$

(iv) $\displaystyle \sum_{Q_j \cap E \neq \emptyset} \left(\frac{r_j}{R_j(y)}\right)^n < \infty$, *where r_j is the diameter of Q_j and $R_j(y) = \mathrm{dist}(Q_j, y)$.*

(v) *There is a measure μ supported on ∂D such that $K(\cdot, \mu) \not\equiv \infty$ and*

$$\lim_{\substack{x \to y \\ x \in E}} \frac{K(x, \mu)}{K(x, y)} = \infty.$$

8. Fine limit approach to the Nagel-Stein boundary limit theorem

8.1. Introduction. The purpose of this section is to discuss some theorems in [11]. Let us first recall the Fatou-Naïm-Doob theorem or the minimal fine limit theorem. This theorem can be stated in a very general framework. Let D be a domain with the Green function $G(x, y)$. We consider the ratio of the Green functions

$$K(x, y) = K_y(x) = \frac{G(x, y)}{G(x_0, y)},$$

where $x_0 \in D$ is a fixed reference point. Suppose $\{y_j\} \subset D$ without accumulation points in D. Since $K(x_0, y_j) = 1$ for all j, the sequence $\{K(\cdot, y_j)\}$ is locally uniformly bounded so that we can choose a subsequence converging to a harmonic function h^*. In this situation, we regard that $\{y_j\}$ defines the boundary point y^* by $h^* = K(\cdot, y^*)$. The totality of such y^* forms an ideal boundary of D, the Martin boundary Δ. The compactification $D \cup \Delta$ is denoted by $\widehat{D}$ and called the Martin compactification. We say that a positive harmonic function h is minimal, if $0 \leq u \leq h$ for a harmonic function u, implies $u = ch$ with some constant c. We write $\Delta_1 = \{y \in \Delta : K_y \text{ is minimal}\}$

and $\Delta_0 = \Delta \setminus \Delta_1$. We call Δ_1 the minimal boundary. The famous Martin representation theorem says that for a positive harmonic function h there exists a unique measure μ_h on Δ_1 such that $h = \int_{\Delta_1} K_y d\mu_h(y)$. Combining with the Riesz decomposition theorem, we obtain the Riesz-Martin representation theorem.

THEOREM 8.A. *For a nonnegative superharmonic function u there exist unique measures μ_u on Δ_1 and ν_u on D such that $u = \int_{\Delta_1} K_y d\mu_u(y) + \int_D K_y d\nu_u(y)$.*

Remark. If D is smooth, then $\widehat{D}$ and Δ_1 are homeomorphic to $\overline{D}$ and ∂D, respectively; $\Delta_0 = \emptyset$. In fact, these hold for a Lipschitz domain ([37]) and an NTA domain ([38]).

DEFINITION. We say that $E \subset D$ is *minimally thin* at $y \in \Delta_1$ if $\widehat{R}^E_{K_y} \neq K_y$. We say that the function f has *minimal fine limit* α at y, denoted by mf $\lim_{x \to y} f(x) = \alpha$, if there is a set E minimally thin at y such that $\lim_{x \to y, x \notin E} f(x) = \alpha$.

THEOREM 8.B. *(Naïm-Doob) Let h be a positive harmonic function with the Martin representing measure μ_h. If u is a nonnegative superharmonic function with the Martin representing measure μ_u for its harmonic part, then*

$$\exists \, \mathrm{mf} \lim_{x \to y} \frac{u(x)}{h(x)} = \frac{d\mu_u}{d\mu_h}(y) \quad \text{for } \mu_h\text{-a.e. } y \in \Delta_1.$$

From Theorem 8.B we can easily deduce the classical Fatou theorem, which asserts that the existence of nontangential boundary values of a positive harmonic function in the unit disk (or in a general smooth domain) at almost all boundary points. The above general results have the following interpretation for a half space. Let $D = \{P = (X, y) : y > 0\}$, where $X = (x_1, \ldots, x_{n-1})$. Then $\widehat{D}$ and Δ_1 are homeomorphic to $\overline{D} \cup \{\infty\}$ and $\partial D \cup \{\infty\}$, respectively; $\Delta_0 = \emptyset$. Observe that y is a positive harmonic function in D. For a bounded set $E \subset D$ the regularized reduced function $\widehat{R}^E_y$ is a Green potential $G\mu_E$. Define the Green energy of E by $\gamma(E) = \int G\mu_E d\mu_E$. We have the following *Wiener criterion* ([42]): E is minimally thin at 0 if and only if

$$\sum 2^{jn} \gamma(E \cap I_j) < \infty,$$

where $I_j = B(0, 2^{-j}) \setminus B(0, 2^{-j-1})$. The following estimate is known: $|E| \le A\gamma(E)$ for a measurable set E; if E is a set of the form $E = \bigcup_{(X,y) \in F} B((X, y), cy)$ with $0 < c < 1$, then $\gamma(E) \approx |E|$ (Corollary 7.4.5, [24], [59]).

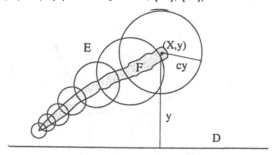

FIGURE 8.1

By the above estimates we have the following integral criterion for minimal thinness (Corollary 7.4.6).

PROPOSITION 8.1.1. *If a measurable set E is minimally thin at 0, then*

$$\int_E |X|^{-n} dX dy < \infty.$$

Moreover, if E is of the form $\{(X, y) : 0 < y < \varphi(|X|)\}$ with nondecreasing function φ, then E is minimally thin at 0 if and only if

$$\int_0 \frac{\varphi(t)}{t^2} dt < \infty.$$

For a further extension see Gardiner [32]. As a simple example $E_a = \{(X, y) : 0 < y < |X|^a\}$ is minimally thin at 0 if and only if $a > 1$. This implies that the minimal fine limit theorem gives no information for tangential behavior through E_a for $a > 1$. Thus the minimal fine limit theorem is irrelevant to tangential boundary behavior.

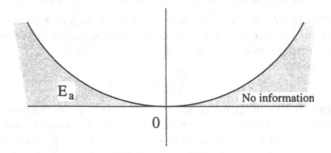

FIGURE 8.2

Surprisingly enough, Nagel and Stein [56] showed that for any given sequence of points in D converging to 0 there is a subdomain Ω of D with $\overline{\Omega} \cap \partial D = \{0\}$ such that Ω contains a subsequence of the given sequence of points; and every positive harmonic function u has boundary limit along $\Omega + X$ for almost all $X \in \partial D$. To be more precise, we denote by $|E|$ the $(n-1)$-dimensional Lebesgue measure of a subset $E \subset \mathbb{R}^{n-1}$. Their domain satisfies the following two conditions:

(i) $|\Omega(y)| \leq Ay^{n-1}$ with $\Omega(y) = \{X : (X, y) \in \Omega\}$ (Cross section condition);
(ii) there is $\alpha > 0$ such that $(X_1, y_1) \in \Omega$ and $|X - X_1| < \alpha(y - y_1) \Longrightarrow (X, y) \in \Omega$ (Cone condition).

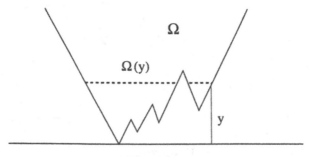

FIGURE 8.3

For simplicity the above two conditions are abbreviated to (NS).

THEOREM 8.C. *Let Ω be a domain in D with $\overline{\Omega} \cap \partial D = \{0\}$. If Ω satisfies (NS), then every positive harmonic function h has boundary limit along $\Omega + X$ for almost all $X \in \partial D$.*

Obviously, a nontangential cone with vertex at the origin satisfies (NS). However, the converse is not necessarily true. In fact, there is Ω satisfying (NS) which contains a sequence of points with prescribed tangency ([56, Lemma 9]). Therefore, Theorem 8.C does not follow from Theorem 8.B. So, it may be natural to ask if we can find a fine limit theorem which yields the Nagel-Stein theorem. In [11] this questions was studied. We define the following Hausdorff type outer measure. For simplicity we write X for a boundary point $(X, 0) \in \partial D$. For a set E we define

$$\Lambda(E) = \inf\{\sum_i r_i^{n-1} : E \subset \bigcup_i B(X_i, r_i), X_i \in \partial D\}.$$

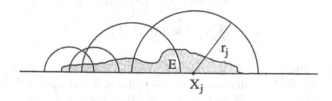

FIGURE 8.4

One should note that a point (X, y) has positive Λ measure unless it lies on ∂D. In fact, $\Lambda(\{(X, y)\}) = y^{n-1}$.

DEFINITION. For $t > 0$ let $D_t = \{(X, y) : 0 < y < t\}$. We say that $E \subset D$ is thin at ∂D with respect to measure if

$$\lim_{t \to 0} \Lambda(E \cap D_t) = 0.$$

We emphasize that a set thin at ∂D with respect to measure is not thin at a particular boundary point. The relationship among this thinness, minimal thinness and ordinary thinness will be given later. Our fine limit type theorem is as follows.

THEOREM 8.1.1. *Let $\Omega \subset D$ and suppose $\overline{\Omega} \cap \partial D = \{0\}$. If h is a positive harmonic function on D, then there exists a set E thin at ∂D with respect to measure such that for almost all $X \in \partial D$, $h(P)$ approaches a limit as P approaches X through $(\Omega + X) \setminus E$.*

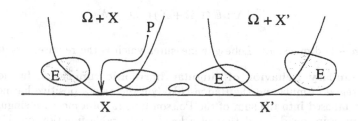

FIGURE 8.5

The proof of Theorem 8.1.1 requires several steps and will be given later. We note that in Theorem 8.1.1 the approach region Ω does not need satisfy (NS). If it satisfies (NS), then $\Omega + X$ essentially avoids the exceptional set E. Thus Theorem 8.C is recaptured by our fine limit theorem.

THEOREM 8.1.2. *Let Ω satisfy (NS). If E is thin at ∂D with respect to measure, then, for almost all $X \in \partial D$, $\Omega + X$ lies eventually outside E, i.e., there is $t = t_X > 0$ such that $E \cap (\Omega + X) \cap D_t = \emptyset$.*

PROOF. Suppose E is thin at ∂D with respect to measure. If $E \cap D_t = \emptyset$, then there is nothing to prove. Suppose $E \cap D_t \neq \emptyset$. By definition there are $X_j \in \partial D$ and $r_j > 0$ such that

$$E \cap D_t \subset \bigcup_j B(X_j, r_j),$$

$$\sum_j r_j^{n-1} < 2\Lambda(E \cap D_t).$$

Let $F_j = \{X : B(X_j, r_j) \cap (\Omega + X) \neq \emptyset\}$. We claim that $F_j \subset B'(X_j, r_j) - \Omega(r_j)$, where in general $B'(X, r)$ denotes the $(n-1)$-dimensional open ball in $\mathbb{R}^{n-1}$ with center at X and radius r. In fact, if $X \in F_j$, then there is $(X_0, y_0) \in \Omega$ such that $(X_0 + X, y_0) \in B(X_j, r_j)$, whence $y_0 \leq r_j$ and $X \in B'(X_j, r_j) - X_0 \subset B'(X_j, r_j) - \Omega(y_0)$. By definition we have the monotonicity of $\Omega(y)$ and so $F_j \subset B'(X_j, r_j) - \Omega(r_j)$. In view of [56, Lemma 1 (d)] we find $\Xi_1, \ldots, \Xi_{N_0} \in \Omega(r_j)$ such that

$$\Omega(r_j) \subset \bigcup_{i=1}^{N_0} B'(\Xi_i, r_j),$$

where N_0 is a number depending only on Ω. Therefore,

$$F_j \subset \bigcup_{i=1}^{N_0} B'(X_j, r_j) - B'(\Xi_i, r_j).$$

The right hand side is the union of balls of radius $2r_j$ and hence $|F_j| \leq A r_j^{n-1}$. Since $\{X : E \cap (\Omega + X) \cap D_t \neq \emptyset\} \subset \bigcup_j F_j$, it follows that

$$|\{X : E \cap (\Omega + X) \cap D_t \neq \emptyset\}| \leq A \sum_j r_j^{n-1} \leq 2A\Lambda(E \cap D_t).$$

By definition the right hand side tends to 0 as $t \to 0$, whence

$$\bigcap_{t>0} \{X : E \cap (\Omega + X) \cap D_t \neq \emptyset\}$$

has zero $(n-1)$-dimensional Lebesgue measure, which is the required result. □

8.2. Boundary behavior of singular harmonic functions. In view of the Poisson integral formula and the Radon-Nikodym theorem a positive harmonic function is decomposed into the sum of the Poisson integral of a measure singular to the Lebesgue measure on ∂D and that of a locally integrable function on ∂D. In this section we consider the boundary behaviour of a singular harmonic function, i.e. a Poisson integral of a measure singular with respect to the Lebesgue measure on ∂D.

THEOREM 8.2.1. *If $h = PI(\mu)$ is the Poisson integral of a finite measure μ singular with respect to the Lebesgue measure on ∂D, then $h(P)$ tends uniformly to 0 as P approaches ∂D outside a set E thin at ∂D with respect to measure, i.e.*

$$\lim_{t \to 0} \left(\sup_{D_t \setminus E} h \right) = 0.$$

We begin with the following weak type estimate. This estimate will also be used in the next section.

LEMMA 8.2.1. *If $h = PI(\mu)$ is the Poisson integral of a finite measure μ, then for each $\varepsilon > 0$*

$$\Lambda(\{P : h(P) > \varepsilon\}) \leq \frac{A}{\varepsilon} \|\mu\|.$$

PROOF. Let $\mathcal{M}\mu(X, r) = \sup_{t > r} \dfrac{\mu(B(X, r))}{r^{n-1}}$. This is the truncated maximal function. We claim that

$$(8.2.1) \qquad h(P) \leq A\mathcal{M}\mu(X, y) \qquad \text{for } P = (X, y).$$

In fact, we write

$$h(P) = PI(\mu)(P) = A \int \frac{y}{|P - Z|^n} d\mu(Z)$$

$$= A \left(\int_{|X-Z| \leq y} \frac{y}{|P - Z|^n} d\mu(Z) + \sum_{j=1}^{\infty} \int_{2^j y < |X-Z| \leq 2^{j+1} y} \frac{y}{|P - Z|^n} d\mu(Z) \right).$$

Observe

$$\int_{|X-Z| \leq y} \frac{y}{|P - Z|^n} d\mu(Z) \leq \frac{y}{y^n} \mu(B(X, y)) \leq \mathcal{M}\mu(X, y).$$

Each integral in the summation is estimated by

$$\int_{2^j y < |X-Z| \leq 2^{j+1} y} \frac{y}{|P - Z|^n} d\mu(Z) \leq \frac{y}{(2^j y)^n} \mu(B(X, 2^{j+1} y)) \leq 2^{n-j} \mathcal{M}\mu(X, y).$$

Hence (8.2.1) follows.

Let $E = \{P : h(P) > \varepsilon\}$. Then (8.2.1) implies that if $P = (X, y) \in E$, then $\mathcal{M}\mu(X, y) \geq A\varepsilon$. In other words, at each point $P = (X, y) \in E$ there is $r = r(P) \geq y$ such that

$$(8.2.2) \qquad r^{n-1} \leq \frac{A}{\varepsilon} \mu(B(X, r)).$$

We may assume that μ is a finite measure and hence $r = r(P)$ is uniformly bounded. Thus the covering lemma (see e.g. [70, Theorem 1.3.5]) is applicable. From the trivial covering $\bigcup_{P \in E} B(P, 2r(P))$ we can extract countably many balls such that

$$E \subset \bigcup_j B(P_j, 2r_j), \text{ where } P_j = (X_j, y_j) \text{ and } r_j = r(P_j) \geq y_j,$$

the multiplicity of $B(P_j, 2r_j)$ is bounded.

Since $B(X_j, r_j) \subset B(P_j, 2r_j) \subset B(X_j, 3r_j)$, it follows that

$$E \subset \bigcup_j B(X_j, 3r_j),$$

the multiplicity of $B(X_j, r_j)$ is bounded.

By (8.2.2) we have

$$\sum r_j^{n-1} \le \frac{A}{\varepsilon} \sum \mu(B(X_j, r_j)) \le \frac{A}{\varepsilon}\|\mu\|.$$

Hence

$$\Lambda(E) \le \frac{A}{\varepsilon}\|\mu\|.$$

The lemma follows. $\square$

The singularity of a measure μ is used in the following lemma.

LEMMA 8.2.2. *If $h = PI(\mu)$ is the Poisson integral of a finite measure μ singular with respect to the Lebesgue measure on ∂D, then for each $\varepsilon > 0$, the set $\{P : h(P) > \varepsilon\}$ is thin at ∂D with respect to measure. That is,*

$$\lim_{t \to 0} \Lambda(\{P : h(P) > \varepsilon\} \cap D_t) = 0.$$

PROOF. Take an arbitrary positive number η. Since μ is singular, we find $X_i \in \partial D$ and $r_i \downarrow 0$ such that

$$\mu(\partial D \setminus \bigcup_i B(X_i, r_i)) = 0,$$

$$\sum_i r_i^{n-1} < \eta.$$

Let μ_j be the restriction of μ on $\bigcup_{i=1}^j B(X_i, r_i)$ and let $\tilde{\mu}_j = \mu - \mu_j$. If $P = (X, y) \in D_t \setminus \bigcup_{i=1}^j B(X_i, 2r_i)$, then

$$PI(\mu_j)(P) \le A \int_{\bigcup_{i=1}^j B(X_i, r_i)} \frac{y}{|P - X'|^n} d\mu(X')$$

$$\le A \int_{\bigcup_{i=1}^j B(X_i, r_i)} \frac{y}{r_j^n} d\mu(X')$$

$$\le A \frac{y}{r_j^n}\|\mu_j\| \le A \frac{t}{r_j^n}\|\mu\|,$$

where we used the fact that $r_1 \ge r_2 \ge \cdots \ge r_j$ in the second inequality. Since $\|\tilde{\mu}_j\| \to 0$ as $j \to \infty$, it follows from Lemma 8.2.1 that there is j such that

$$\Lambda(\{PI(\tilde{\mu}_j) > \frac{\varepsilon}{2}\}) < \eta.$$

For this j we take $t > 0$ so small that $A\frac{t}{r_j^n}\|\mu\| < \varepsilon/2$. Then

$$\Lambda(\{h > \varepsilon\} \cap D_t) \le \Lambda\left(\{h > \varepsilon\} \cap D_t \setminus (\bigcup_{i=1}^j B(X_i, 2r_i))\right) + \Lambda(\bigcup_{i=1}^j B(X_i, 2r_i))$$

$$\le \Lambda(\{PI(\tilde{\mu}_j) > \frac{\varepsilon}{2}\}) + A \sum_i r_i^{n-1}$$

$$\le A\eta.$$

The lemma follows. $\square$

PROOF OF THEOREM 8.2.1. Take $\varepsilon_i \downarrow 0$. From Lemma 8.2.2 we find $t_i \downarrow 0$ such that

$$\sum_i \Lambda(\{h > \varepsilon_i\} \cap D_{t_i}) < \infty.$$

Let $E = \bigcup_i E_i$ with $E_i = \{h > \varepsilon_i\} \cap (D_{t_i} \setminus D_{t_{i+1}})$.

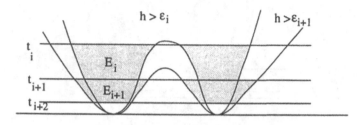

FIGURE 8.6

Then E is thin at ∂D with respect to measure since

$$\Lambda(E \cap D_{t_k}) \le \sum_{i=k}^{\infty} \Lambda(E_i) \le \sum_{i=k}^{\infty} \Lambda(\{h > \varepsilon_i\} \cap D_{t_i}) \to 0$$

as $k \to \infty$. By definition

$$\sup_{D_{t_k} \setminus E} h = \sup_{i \ge k} \left(\sup_{(D_{t_i} \setminus D_{t_{i+1}}) \setminus E_i} h \right) \le \sup_{i \ge k} \varepsilon_i = \varepsilon_k \to 0$$

as $k \to \infty$. Thus the theorem is proved. $\square$

8.3. Proof of Theorem 8.1.1. The proof in [11] used oscillation and was rather complicated. Below we give a clearer proof based on the Lusin theorem. This proof is suggested by Alexander Borichev. Let us begin by recalling the Lusin theorem (see e.g. [41, Theorem 3.3]).

THEOREM 8.D. (Lusin Theorem) *Let f be an integrable function on ∂D with compact support. Then for each $\varepsilon > 0$ there is a continuous function g on ∂D with compact support such that f and g coincide outside a set of measure less than ε and $\|f - g\|_1 < \varepsilon$.*

We remark that in [41, Theorem 3.3] the Lusin theorem is stated for more general functions. For an integrable function we can easily modify g so that the additional assumption $\|f - g\|_1 < \varepsilon$ is satisfied.

In order to prove Theorem 8.1.1, we may assume, without loss of generality, that Ω is bounded. Hence $\overline{\Omega} \cap \partial D = \{0\}$ implies that for each $r > 0$ there is $t > 0$ such that $\Omega \cap \{0 < y < t\} \subset B(0, r)$, or equivalently $\Omega \setminus B(0, r) \subset \{y \ge t\}$. Suppose $r_j \downarrow 0$ is given. Then, we can choose $t_j \downarrow 0$ so that

$$(8.3.1) \qquad\qquad \Omega \cap \{0 < y < t_j\} \subset B(0, r_{j+1}),$$

or equivalently $\Omega \setminus B(0, r_{j+1}) \subset \{y \ge t_j\}$. Observe that

$$\Omega \cap B(0, r_i) = \bigcup_{j=i}^{\infty} \Omega \cap B(0, r_j) \setminus B(0, r_{j+1}) \subset \bigcup_{j=i}^{\infty} B(0, r_j) \cap \{y \ge t_j\}.$$

By translation we have for every $X \in \partial D$

$$(8.3.2) \qquad (\Omega + X) \cap B(X, r_i) \subset \bigcup_{j=i}^{\infty} B(X, r_j) \cap \{y \geq t_j\}.$$

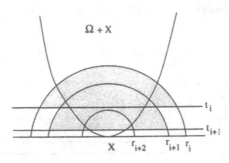

FIGURE 8.7

PROOF OF THEOREM 8.1.1. In view of the Radon-Nikodym theorem and Theorem 8.2.1, we may assume that h is the Poisson integral of a locally integrable function f on ∂D. Moreover, we may assume that f has a compact support.

Let $\varepsilon_j \downarrow 0$ and $\eta_j \downarrow 0$. Suppose

$$\sum_j \eta_j < \infty.$$

By the Lusin theorem we can find continuous functions g_j of compact support and sets $F_j \subset \partial D$ such that

$$(8.3.3) \qquad f(X) = g_j(X) \quad \text{for } X \notin F_j \text{ with } |F_j| < \eta_j,$$
$$(8.3.4) \qquad \|f - g_j\|_1 < \eta_j \varepsilon_j.$$

Note that g_j and $PI(g_j)$ are uniformly continuous and $PI(g_j)$ has the boundary values g_j. Hence we can find $r_j \downarrow 0$ such that

$$(8.3.5) \qquad |PI(g_j) - g_j(X)| < \varepsilon_j \quad \text{on } B(X, r_j) \cap D$$

uniformly for all $X \in \partial D$. By Lemma 8.2.1 and (8.3.4) we have

$$(8.3.6) \qquad \Lambda(E_j) \leq A \frac{\|f - g_j\|_1}{\varepsilon_j} \leq A\eta_j,$$

where $E_j = \{P : |PI(f)(P) - PI(g_j)(P)| \geq \varepsilon_j\}$. We have from (8.3.3) and (8.3.5)

$$|PI(f) - f(X)| < 2\varepsilon_j \quad \text{on } B(X, r_j) \cap D \setminus E_j$$

for $X \in \partial D \setminus F_j$. Let $t_j \downarrow 0$ be as in (8.3.1), $E_j' = E_j \cap \{y \geq t_j\}$ and $E = \bigcup_j E_j'$. Then

$$(8.3.7) \qquad |PI(f) - f(X)| < 2\varepsilon_j \quad \text{on } B(X, r_j) \cap \{y \geq t_j\} \setminus E$$

for $X \in \partial D \setminus F_j$.

Now the theorem follows if we prove the following statements:

  (i) E is thin at ∂D with respect to measure.
  (ii) $|F^*| = 0$, where $F^* = \bigcap_i \bigcup_{j \geq i} F_j$.
  (iii) $\lim_{P \to X, P \in (\Omega + X) \setminus E} PI(f)(P) = f(X)$ for $X \in \partial D \setminus F^*$.

From (8.3.6) we see that

$$\Lambda(E \cap D_{t_i}) = \Lambda(\bigcup_{j \geq i} E'_j) \leq \sum_{j \geq i} \Lambda(E'_j) \leq \sum_{j \geq i} \Lambda(E_j) \leq A \sum_{j \geq i} \eta_j \downarrow 0 \text{ as } i \uparrow \infty.$$

Thus E is thin at ∂D with respect to measure. From (8.3.3) we see that

$$|F^*| \leq \sum_{j \geq i} |F_j| \leq \sum_{j \geq i} \eta_j \downarrow 0 \text{ as } i \uparrow \infty.$$

Hence $|F^*| = 0$. Finally we prove (iii). Let $X \in \partial D \setminus F^*$. Then there is $j_0 = j_0(X)$ such that if $j \geq j_0$, then $X \in \partial D \setminus F_j$ and (8.3.7) holds. Let $i > j_0$. We observe from (8.3.2) that

$$(\Omega + X) \cap B(X, r_i) \setminus E \subset \bigcup_{j \geq i} B(X, r_j) \cap \{y \geq t_j\} \setminus E.$$

Hence (8.3.7) implies

$$|PI(f) - f(X)| < 2 \sup_{j \geq i} \varepsilon_j = 2\varepsilon_i \quad \text{on } (\Omega + X) \cap B(X, r_i) \setminus E.$$

This implies (iii). The theorem is proved. $\square$

8.4. Sharpness of Theorem 8.1.1 and Theorem 8.2.1. Let us show that the size of the exceptional sets in Theorem 8.1.1 and Theorem 8.2.1 is best possible for the boundary behavior of harmonic functions.

THEOREM 8.4.1. *Suppose E is thin at ∂D with respect to measure. Then*
(i) there exists a nonnegative integrable function f on ∂D such that the Poisson integral $PI(f)(P)$ tends uniformly to ∞ as P approaches ∂D through E;
(ii) there exists a nonnegative singular measure μ on ∂D such that the Poisson integral $PI(\mu)(P)$ tends uniformly to ∞ as P approaches ∂D through E.

For the proof of (i) we shall use the fact that

(8.4.1) $$PI(\chi_{B'(X, 2r)}) \geq A \quad \text{on } B(X, r) \cap D.$$

The proof of (ii) is more subtle. The following lemma corresponds to (8.4.1).

LEMMA 8.4.1. *Let $X \in \partial D$ and $0 < t < r$. Then there exists a measure $\mu_{X,r,t}$ such that*
(i) *$\mu_{X,r,t}$ consists of point masses at $A(r/t)^{n-1}$ many points in $B'(X,r)$ of magnitude t^{n-1};*
(ii) *$PI(\mu_{X,r,t}) \geq A$ on $B(X,r) \setminus D_t$.*

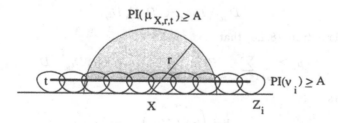

FIGURE 8.8

PROOF. We can choose $A(r/t)^{n-1}$ many points $Z_i \in B'(X, 2r)$ such that $B'(X, 2r) \subset \bigcup B'(Z_i, t)$. Observe that $PI(\nu_i) \geq A$ on $B'(Z_i, t) \times \{t\}$, where ν_i is the point mass at $Z_i \in \partial D$ of magnitude t^{n-1}. Let $\mu_{X,r,t} = \sum_i \nu_i$. Then $h = PI(\mu_{X,r,t}) \geq A$ on $B'(X, 2r) \times \{t\}$. By the maximum principle and translation we obtain from (8.4.1) that $h \geq A$ on $B(X, r) \setminus D_t$. The lemma follows. $\square$

PROOF OF THEOREM 8.4.1. By definition we can find a decreasing sequence t_j such that

$$\sum_{j=1}^{\infty} \Lambda(E_j) < \infty,$$

where $E_j = E \cap D_{t_j}$. Then we find $X_{j,i} \in \partial D$ and $r_{j,i} > 0$ such that

(8.4.2) $$E_j \subset \bigcup_i B(X_{j,i}, r_{j,i}),$$

(8.4.3) $$\sum_{j,i} r_{j,i}^{n-1} < \infty.$$

In order to prove (i), we let $f_j = \sum_i \chi_{B'(X_{j,i}, 2r_{j,i})}$ and $f = \sum_j f_j$. Then f is integrable by (8.4.3) and $PI(f_j) \geq A$ on E_j by (8.4.1) and (8.4.2). Hence

$$PI(f) \geq \sum_{j=1}^{j_0} PI(f_j) \geq A j_0 \quad \text{on } E_{j_0}.$$

This implies that

$$\lim_{j \to \infty} \left(\inf_{E_j} PI(f) \right) = \infty.$$

Thus (i) follows.

Let us now prove (ii). Let $\mu_{j,i}$ be the measure obtained in Lemma 8.4.1 with $X = X_{j,i}$, $r = r_{j,i}$ and $t = t_{2j}$. Let $\mu_j = \sum_i' \mu_{j,i}$, where $\sum_i'$ stands for the finite summation over i with $r_{j,i} > t_{2j}$. We have from (8.4.2) and Lemma 8.4.1

(8.4.4) $$PI(\mu_j) \geq A \quad \text{on } E_j \setminus D_{t_{2j}} = E \cap D_{t_j} \setminus D_{t_{2j}};$$

(8.4.5) $$\|\mu_j\| \leq A \sum_i r_{j,i}^{n-1}.$$

Let $\mu = \sum_j \mu_j$. Then, by (8.4.3), μ is a finite singular measure concentrated on at most countably many points. Moreover, observe that if $j_0 + 1 \leq 2j \leq 2j_0$, then $t_{2j} \leq t_{j_0+1} < t_{j_0} \leq t_j$, so that

$$D_{t_{j_0}} \setminus D_{t_{j_0+1}} \subset D_{t_{2j}} \setminus D_{t_j}.$$

Hence it follows from (8.4.4) that

$$PI(\mu) \geq \sum_{j_0+1 \leq 2j \leq 2j_0} PI(\mu_j) \geq A j_0 \quad \text{on } E \cap D_{t_{j_0}} \setminus D_{t_{j_0+1}},$$

which implies

$$\lim_{j \to \infty} \left(\inf_{E_j} PI(\mu) \right) = \infty.$$

Thus (ii) follows. $\square$

Remark. (i) Let $G = \bigcup_j G_j$ be the union of 'grids' G_j in [9, §2]. Then $G \cap B(0, R)$ is thin at ∂D with respect to measure for any $R > 0$. Moreover, $\overline{G} \cap \partial D = \partial D$. We constructed a harmonic function u such that $|u| \leq 1$ and $\operatorname*{osc}_G(h; X, r) = 2$ for all $r > 0$ and all $X \in \partial D$ ([9, Theorem 1]). This shows again that the size of the exceptional set in Theorem 8.2.1 is best possible.

(ii) Ahern [6] studied the boundary behavior of the Poisson integral of a singular measure. He described the $n - 1$ dimensional measure of the intersection of the hyperplane $\{(X, y) : X \in \mathbb{R}^{n-1}\}$ and the set of points at which the Poisson integral is greater than 1. It may be interesting to compare his results and our Theorem 8.1.1 and Theorem 8.4.1 (ii).

8.5. Necessity of an approach region. Note that we cannot remove Ω. This fact is suggested by Alexander Borichev. In fact we have

THEOREM 8.5.1. *Let f be an integrable function and let $X_0 \in \partial D$. Suppose that there is a set E thin at ∂D with respect to measure such that*

$$(8.5.1) \qquad \lim_{P \to X_0, P \notin E} PI(f)(P) = \alpha.$$

Then

$$(8.5.2) \qquad \operatorname*{ess\,lim}_{X \to X_0} f(X) = \alpha.$$

Remark. (i) (8.5.2) $\iff$ for every $\varepsilon > 0$ there is $r > 0$ such that $|f(X) - \alpha| < \varepsilon$ a.e. for $X \in B'(X_0, r)$.

(ii) If (8.5.2) holds, then (8.5.1) holds with $E = \emptyset$.

(iii) There is a bounded integrable function f such that $\operatorname{ess\,lim}_{X \to X_0} f(X)$ does not exist at every $X_0 \in \partial D$ (For a construction see e.g. [9, Theorem 1].)

For a subset E of D the set

$$\{X : \text{for any } t > 0 \text{ there is } y, \ 0 < y < t \text{ such that } (X, y) \in E\}$$

is called the *essential projection* of E. As an easy corollary to Theorem 8.1.2, we see that the essential projection of a set thin at ∂D with respect to measure is of measure 0. Hence Theorem 8.5.1 follows from the next theorem.

THEOREM 8.5.2. *Let f be an integrable function and let $X_0 \in \partial D$. Suppose that there is a set E such that the essential projection of E is of measure 0 and (8.5.1) holds. Then (8.5.2) holds.*

PROOF. We prove the theorem by contradiction. Suppose (8.5.2) does not hold. Then there is $\varepsilon_0 > 0$ such that

$$(8.5.3) \qquad |\{X \in B'(X_0, r) : |f(X) - \alpha| > 2\varepsilon_0\}| > 0$$

for all $r > 0$. By (8.5.1) we see that there is $r_0 > 0$ such that

$$(8.5.4) \qquad |PI(f)(P) - \alpha| < \varepsilon_0 \quad \text{for } P \in B(X_0, r_0) \cap D \setminus E.$$

Let us consider the boundary values of $PI(f)$ on $B'(X_0, r_0)$. It is well known that $PI(f)$ has nontangential boundary values f a.e. on ∂D. Hence, by (8.5.3), we can find a set $F \subset B'(X_0, r_0)$ such that $|F| > 0$ and

$$\left| \operatorname*{nt\,lim}_{P \to X} PI(f)(P) - \alpha \right| > 2\varepsilon_0 \quad \text{for } X \in F,$$

where nt $\lim_{P\to X} PI(f)(P)$ means the nontangential limit of $PI(f)$ at X. In particular, there is a positive function $t(X)$ on F such that

$$|PI(f)(X,y) - \alpha| > \varepsilon_0 \quad 0 < y < t(X),$$

or, in other words

$$|PI(f)(P) - \alpha| > \varepsilon_0 \quad \text{on } T = \bigcup_{X\in F} I(X),$$

where $I(X) = \{X\} \times (0, t(X))$. By (8.5.4) we see that $T \cap B(X_0, r_0) \subset E$. Obviously, F is the essential projection of $T \cap B(X_0, r_0)$. Hence the essential projection of E has positive measure. This is a contradiction. $\square$

8.6. Further results. Here we collect further results related to sets thin at boundary with respect to measure. No proofs will be given. For details we refer to [11]. First we introduce a different notion of thinness. This notion was essentially given by Rippon [58]. Let $\{Q_j\}$ be the Whitney decomposition of D (see [62, Chapter VI]) and let $t_j = \text{dist}(Q_j, \partial D)$. We observe that the diameter r_j of Q_j is comparable to t_j. By cap(E) we denote the logarithmic capacity of E if $n = 2$, and the Newtonian capacity of E if $n \geq 3$.

DEFINITION 8.6.1. We say that $E \subset D$ is thin at ∂D with respect to capacity if

$$\sum_k t_j \left(\log \frac{4t_j}{\text{cap}(E \cap Q_j)} \right)^{-1} < \infty \quad \text{for } n = 2,$$

$$\sum_k t_j \, \text{cap}(E \cap Q_j) < \infty \qquad \text{for } n \geq 3.$$

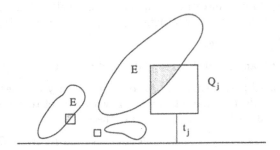

FIGURE 8.9

The above thinness is needed in the study of boundary behavior of Green potentials.

THEOREM 8.6.1. *Let u be a Green potential. Then there exist a set E thin at ∂D with respect to capacity and a set F thin at ∂D with respect to measure such that $u(P)$ converges to 0 as P approaches ∂D outside $E \cup F$.*

Our thinness is related to minimal thinnes and ordinary thinness in the following fashion.

THEOREM 8.6.2. *(i) If E is thin at ∂D with respect to capacity, then E is minimally thin at almost all $X \in \partial D$.*

(ii) If F is thin at ∂D with respect to measure, then F is minimally thin at almost all $X \in \partial D$.

(iii) If a bounded set T is minimally thin at almost all $X \in \partial D$, then T can be decomposed into the union of sets thin at ∂D with respect to capacity and with respect to measure.

(iv) Let Ω satisfy (NS). If E is thin at ∂D with respect to capacity, then $E \cap (\Omega + X)$ is thin (for $n \geq 3$) or nearly thin (for $n = 2$, cf. [34, Chapter 7]) at X for almost all $X \in \partial D$.

Let $\Omega \subset D$ and suppose $\overline{\Omega} \cap \partial D = \{0\}$. We say that f has Ω-fine (resp. Ω-near fine) limit α at $X \in \partial D$ if $f(P)$ tends to α as P approaches X through $(\Omega + X) \setminus E$ with some set E thin (resp. nearly thin) at X. From Theorem 8.A, Theorem 8.C, Theorem 8.1.1, Theorem 8.1.2 and Theorem 8.6.2, we obtain the following result, which may be considered to be a superharmonic version of Theorem 8.C.

THEOREM 8.6.3. *Let Ω satisfy (NS). Then every nonnegative superharmonic function on D has Ω-fine limit (for $n \geq 3$) or Ω-near fine limit (for $n = 2$) at almost all $X \in \partial D$.*

Finally we note that nontangential cones are related to our thin sets in the following manner.

THEOREM 8.6.4. *Let $\Gamma(X_0, a, t)$ be the truncated nontangential cone*

$$\{(X, y) : |X - X_0| < ay, \ 0 < y < t\}$$

with vertex $X_0 \in \partial D$, aperture $a > 0$ and height $t > 0$. Suppose $a(X) > 0$ and $t(X) > 0$ for almost all $X \in \partial D$. Then for each $R > 0$ the intersection of $B(0, R)$ and the set

$$D \setminus \bigcup_{X \in \partial D} \Gamma(X, a(X), t(X))$$

is thin at ∂D with respect to measure.

Brelot and Doob [22, Théorème 8] proved that the above set is minimally thin at almost all boundary points. Theorem 8.6.4 is, in a sense, a refinement of their result.

9. Integrability of superharmonic functions

9.1. Integrability for smooth domains. Let D be a domain in $\mathbb{R}^n$ with $n \geq 2$. Throughout this section we let u be a nonnegative superharmonic function on D. First of all let us consider the local integrability of u. Let us take a compact subset K of D. By the Riesz theorem we know that u is decomposed into the sum of a harmonic function and a Green potential. The harmonic function is continuous and therefore L^p-integrable on K for any $p > 0$. On the other hand, the Green function has the singularity of order $2 - n$ for $n \geq 3$ and the logarithmic singularity for $n = 2$. Hence $u \in L^p(K)$ for $0 < p < n/(n - 2)$. This is the local integrability.

The global integrability, or the integrability near the boundary, is more delicate. In early 70's D. H. Armitage first realized this problem. In his two articles ([17], [18]) he dealt with the problem for smooth domains. Let us assume that for each boundary point ξ we can find an interior ball B_i and an exterior ball B_e whose radii are bounded below by some positive constant r. Armitage called such a domain a *domain of bounded curvature*. Obviously, a C^2 domain is of bounded curvature. He showed the following.

THEOREM 9.A. *If D is of bounded curvature, then $u \in L^p(K)$ for $0 < p < n/(n-1)$.*

Since the proof is not so difficult, we give a sketch. We fix a reference point $x_0 \in D$. By $\delta(x)$ we denote the distance $\text{dist}(x, \partial D)$. Let $G(x, y)$ be the Green function for D. From the existence of the exterior ball we obtain the upper estimate

$$G(x_0, y) \leq A\delta(y);$$

from the existence of the interior ball we obtain the lower estimate

$$G(x_0, y) \geq A\delta(y).$$

These estimates hold near the boundary. Here the constant A depends only on the dimension n, the radius r and the fixed point x_0. From these inequalities we can show the following key estimate:

(9.1.1) $$G(x, y) \leq AG(x_0, y)|x - y|^{1-n}$$

Taking the above estimate for granted, we can prove the theorem easily.

PROOF OF THEOREM 9.A. By approximation we may assume that u is the Green potential

$$u(x) = \int_D G(x, y)d\mu(y).$$

Take the exponent p, $1 \leq p < n/(n-1)$. Then, by the Minkowski inequality in the integral form, we get the following inequalities:

$$\left(\int_D u(x)^p dx \right)^{1/p} = \left(\int_D \left(\int_D G(x, y)d\mu(y) \right)^p dx \right)^{1/p}$$

$$\leq \int_D \left(\int_D G(x, y)^p dx \right)^{1/p} d\mu(y)$$

$$\leq A \int_D G(x_0, y) \left(\int_D |x - y|^{p(1-n)} dx \right)^{1/p} d\mu(y)$$

$$\leq A \int_D G(x_0, y)d\mu(y) = Au(x_0).$$

Since u is finite at least one point and x_0 is arbitrary, the L^p norm of u is finite. $\square$

9.2. Integrability for Lipschitz domains. For the succeeding work, we had to wait for a long time. There was no developments of this problem for years. Nearly 17 years later Maeda and Suzuki [46] extended Armitage's result. Namely, they succeeded to treat a Lipschitz domain. Since a Lipschitz domain may have corners and edges, Armitage's argument does not apply to a Lipschitz domain. We need some terminology. We say that a bounded domain D is k-Lipschitz if D and ∂D are given locally by a Lipschitz function whose Lipschitz constant is at most k. A Lipschitz domain is a k-Lipschitz domain for some $k > 0$. We can find an interior cone and an exterior cone with fixed aperture ψ. This aperture ψ is, of course, related to the Lipschitz constant of the graph. Namely, $\cot \psi = k$ for k-Lipschitz domain.

Maeda-Suzuki's argument is based on the following barrier functions. For the inside cone Γ_i we associate the harmonic measure that takes the value 1 on the round boundary. We call this harmonic measure w the inside barrier. Then we see that $w(x)$ is similar to $\delta(x)^\alpha$ in a nontangential part. Let us call this constant α the *maximal order of barrier*. Obviously, the maximal order of barrier depends on the angle ψ and the dimension n. So, we write $\alpha_n(\psi)$. Since Γ_i is inside the domain, this barrier shows how slowly the Green function decays at the boundary.

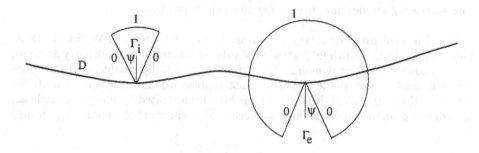

FIGURE 9.1

For the outside cone Γ_e we shall consider the complement in the ball and associate the harmonic measure as illustrated in the above figure. This is the exterior barrier function. The exterior barrier function indicates how quickly the Green function decays at the boundary. The order is related to the angle by $\beta = \alpha_n(\pi - \psi)$. Let us note that $0 < \beta < 1 < \alpha$ and α and β tend to 1 as as $\psi \to \dfrac{\pi}{2}$, or equivalently $k \to 0$.

Now it is easy to see that the existence of the interior cone Γ_i implies the lower estimate

$$(9.2.1) \qquad\qquad G(x_0, y) \geq A\delta(y)^\alpha$$

and the existence of the exterior cone Γ_e implies the upper estimate

$$G(x_0, y) \leq A\delta(y)^\beta.$$

From these estimates we can show the following key estimate.

$$(9.2.2) \qquad\qquad G(x, y) \leq AG(x_0, y)\delta(x)^\beta |x - y|^{2-n-2\alpha}.$$

Let us compare this estimate with the estimate (9.1.1) for a smooth domain. Since $0 < \beta < 1 < \alpha$, our present estimate is worse than (9.1.1). A simple adaptation of the Minkowski inequality does not work well. But the estimate still gives us some information for the integrability. In fact, Maeda–Suzuki used (9.2.2) in an ingenious fashion, and proved

THEOREM 9.B. *There is a positive constant p_k such that if D is a k-Lipschitz domain, then $u \in L^p(D)$ for $0 < p < p_k$.*

Remark. As we have observed in the key estimate, α and β tend to 1 as $k \to 0$. From this fact we can show that $p_k \to n/(n-1)$ as $k \to 0$. For a C^1-domain, if we take a small portion of the boundary, then we can represent it as the k-Lipschitz graph with arbitrarily small k. Hence we obtain the same conclusion as for a smooth domain. Namely, $u \in L^p(D)$ for $0 < p < n/(n-1)$ for a C^1 domain.

Remark. However, we should remark that the constant p_k given by Maeda-Susuki is not sharp. In many cases, $p_k < 1$, and so the most interesting case when every u is integrable is dropped.

For further extensions see Suzuki [63], [64] and [65]. Anyhow, by the remarkable work of Maeda–Suzuki, we have two directions in this subject.

(i) Extend the results to a nasty domain.
(ii) Get the sharp bound of p.

The succeeding studies have been going along these two lines.

9.3. Integrability for nasty domains. In the first direction, W. Smith, D. A. Stegenga and D. C. Ullrich [61] extended Maeda–Suzuki result to more nasty domains, namely, so-called Hölder domains. This terminology is rather confusing. Unlike Lipschitz domains, their Hölder domain is not represented as the graph of a Hölder function. (Recently, graph-Hölder domains have been treated by many probabilists.) In order to define a Hölder domain we need the quasi-hyperbolic metric k_D, defined by

$$k_D(x, y) = \inf \int_\gamma \frac{ds}{\delta(z)},$$

where the infimum is taken over all rectifiable curves γ that connects x and y. We call a domain a *Hölder domain* if there are constants A and A' such that

$$k_D(x, x_0) \le A \log \frac{1}{\delta(x)} + A' \quad \text{for all } x \in D.$$

By definition the quasi-hyperbolic metric is closely related to the Harnack inequality. In fact, the quasi-hyperbolic metric $k_D(x, y)$ is comparable to the length of of the shortest Harnack chain that connect x and y. Hence the Harnack inequality implies that there is a constant $\alpha > 1$, depending only on the dimension, such that

$$\frac{h(x)}{h(y)} \le \exp(\alpha k_D(x, y))$$

for every positive harmonic function h. Stegenga and Ullrich found that the same inequality still holds for a nonnegative superharmonic function u with a different constant α' in a certain weak sense. In particular,

$$u(x) \le u(x_0) \exp(\alpha' k_D(x, x_0)) \quad \text{in some sense.}$$

Hence, the integrability of u is reduced to the integrability of the quasi-hyperbolic metric $k_D(x, x_0)$, especially its exponential integrability.

The exponential integrability for a Hölder domain was established by W. Smith and D. A. Stegenga [60]. They showed that there is a small positive constant $\varepsilon > 0$ such that

$$\int_D \exp(\varepsilon k_D(x, x_0))dx < \infty.$$

Thus we obtain

THEOREM 9.C. *If D is a Hölder domain, then there is a small positive constant $p_D > 0$ such that $u \in L^p(D)$ for $0 < p < p_D$.*

The exact value of p_D is not known. Since it depends on the exponential integrability constant ε, and this ε is obtained by the BMO technique, it seems hopeless to get the sharp value of p_D by their method.

Remark. Somewhat similar BMO technique was used by P. Lindqvist [45] to show the integrability of a positive supersolution of certain nonlinear equations, such as the p-Laplace equation.

9.4. Sharp integrability for plane domains. M. Masumoto ([47], [48], [49]) succeeded to obtain the sharp value of p_k for plane domains. He used a deep argument in function theory. Let us give a rough sketch of his idea. Assume that D is a domain bounded by finitely many closed Jordan curves. Suppose D_0 is a smooth domain and f is a conformal mapping from D_0 onto D. Obviously, if u is superharmonic on D, then $u \circ f$ is superharmonic on D_0. Let λ_D and λ_{D_0} be the Poincaré metrics for D and D_0, respectively. They are related as $\lambda_{D_0}(z) = \lambda_D(f(z))|f'(z)|$ so that

$$\iint_D \lambda_D(Z)^\alpha u(Z) dX dY = \iint_{D_0} \left(\frac{\lambda_{D_0}(z)}{|f'(z)|} \right)^\alpha u(f(z))|f'(z)|^2 dx dy$$

$$= \iint_{D_0} \lambda_{D_0}(z)^\alpha |f'(z)|^{2-\alpha} u(f(z)) dx dy.$$

Poincaré metric is estimated by the ordinary distance: $\lambda_{D_0}(z) \approx \delta_{D_0}(z)^{-1}$ and $\lambda_D(Z) \approx \delta_D(Z)^{-1}$. For the smooth domain D_0 we have the integrability

$$\iint_{D_0} \lambda_{D_0}(z)^\gamma u(z) dx dy \approx \iint_{D_0} \delta_{D_0}(z)^{-\gamma} u(z) dx dy < \infty \quad \text{for } \gamma < 1.$$

Therefore, if $|f'(z)|$ is estimated, then the integrability for D is obtained. In fact, such an estimate was given by F. D. Lesley [43]. Let $0 < \theta < 1$. We say that a domain D satisfies an interior θ-wedge condition if there exists $\rho > 0$ such that, for every boundary point z, an open sector of radius ρ and opening $\pi\theta$ with vertex at z lies in D. He showed that

$$|f'(z)| \le A\lambda_{D_0}(z)^{1-\theta}$$

if D satisfies an interior θ-wedge condition. Hence, if $\alpha + (2-\alpha)(1-\theta) < 1$ or equivalently $\alpha < (2\theta - 1)/\theta$, then

$$\iint_D \delta_D(Z)^{-\alpha} u(Z) dX dY < \infty.$$

Note that each k-Lipschitz domain satisfies an interior θ-wedge condition for $\theta = (2/\pi)\arctan(1/k)$. Hence we have

THEOREM 9.D. *Let D be a k-Lipschitz domain with $0 < k < 1$. Then $u \in L^1(D)$.*

Applying Fubini's theorem, we finally obtain

THEOREM 9.E. *Let D be a k-Lipschitz domain. Then $u \in L^p(D)$ for $0 < p < p_k$ with*

$$p_k = \begin{cases} \dfrac{4\psi}{\pi} & \text{if } 0 < k \le 1, \\ \dfrac{2\psi}{\pi - 2\psi} & \text{if } k > 1, \end{cases}$$

where $k = \cot\psi$. The constant p_k is sharp.

9.5. Sharp integrability for Lipschitz domains. For the higher dimensional case we cannot use the conformal technique. In order to obtain the sharp integrability we need a different device. This is the *coarea formula*.

LEMMA. *Let f be a nonnegative smooth function and ψ a nonnegative Borel measurable function on D. Then we have the following identity:*

$$\int_D \psi(x)|\nabla f(x)| dx = \int_0^\infty dt \int_{\{x \in D: f(x) = t\}} \psi(x) d\sigma(x).$$

This identity means that the volume integral is converted to a double integral, the first one is the surface integral over the level surface, and the second one an 1 dimensional integral. It may be taken as a kind of Fubini theorem. For a proof we refer to [52, pp. 37–39]. Let us apply the coarea formula to obtain the following main identity and two inequalities ([13, Theorem]).

THEOREM 9.5.1. *Let $g(x)$ be the Green function $G(x, x_0)$ with pole at $x_0 \in D$. Let $\varphi(t)$ be a nonnegative Borel measurable function for $t > 0$ and let h, u and v be nonnegative functions on D which are harmonic, superharmonic and subharmonic on D, respectively. Then*

$$\int_D h(x)\varphi(g(x))|\nabla g(x)|^2 dx = c_n h(x_0) \int_0^\infty \varphi(t)dt,$$

$$\int_D u(x)\varphi(g(x))|\nabla g(x)|^2 dx \leq c_n u(x_0) \int_0^\infty \varphi(t)dt,$$

$$\int_D v(x)\varphi(g(x))|\nabla g(x)|^2 dx \geq c_n v(x_0) \int_0^\infty \varphi(t)dt,$$

where $c_2 = 2\pi$ and $c_n = (n-2)\sigma_n$ for $n \geq 3$ and σ_n is the surface measure of a unit sphere.

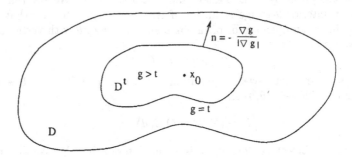

FIGURE 9.2

PROOF. Let D^t be the subdomain given by

$$D^t = \{x \in D : g(x) > t\}.$$

Let $G_t(x, y)$ be the Green function for D^t. Then, we see

$$G_t(x, x_0) = G(x, x_0) - t = g(x) - t,$$

because the Green function is the unique function that has the singularity at x_0 and vanishes on the boundary. The function $g(x) - t$ has these two properties, so it must coincide with the Green function $G_t(x, x_0)$.

Since ∂D^t is smooth, the Green formula shows that the harmonic measure ω_t at x_0 for D^t is given by the following identity

$$c_n \frac{d\omega_t}{d\sigma} = -\frac{\partial G_t(\cdot, x_0)}{\partial n} = -\frac{\partial g}{\partial n} = |\nabla g| \quad \text{on } \partial D^t.$$

The last equality follows from the fact that the outward normal of the level surface $\{g = t\}$ is given by $-\dfrac{\nabla g}{|\nabla g|}$. It should be noted that this calculation does not hold for all t, but it holds for almost all t. This is the Sard theorem. Of course the Sard theorem is used in the proof of the coarea formula.

Now let us prove the theorem for a harmonic function h. We see that the Poisson integral formula becomes

$$h(x_0) = \frac{1}{c_n} \int_{\partial D^t} h |\nabla g| d\sigma.$$

We invoke the coarea formula with $\psi = h\varphi(g)|\nabla g|$, $f = g$. Then

$$\int_D h(x)\varphi(g(x))|\nabla g(x)|^2 dx = \int_D h(x)\varphi(g(x))|\nabla g(x)||\nabla g(x)|dx$$

$$= \int_0^\infty dt \int_{g=t} \varphi(g)h|\nabla g|d\sigma = \int_0^\infty \varphi(t)dt \int_{\partial D^t} h|\nabla g|d\sigma$$

$$= c_n h(x_0) \int_0^\infty \varphi(t)dt.$$

It is easy to treat the two remaining cases. The theorem is proved. $\square$

The above theorem has lots of applications. First of all we note the following elementary inequality:

$$|\nabla g(x)| \le A\frac{g(x)}{\delta(x)} \quad \text{near } \partial D.$$

This is easy. Write the harmonic function g by the Poisson integral and differentiate it under the integral sign. Substitute the inequality to the subharmonic version of Theorem 9.5.1. Then

$$\int_D v(x)\varphi(g(x))|\nabla g(x)|^2 dx \ge c_n v(x_0) \int_0^\infty \varphi(t)dt.$$

This gives the nonintegrability of subharmonic functions.

COROLLARY 9.5.1. *Let φ be a doubling function (i.e. $\varphi(t) \approx \varphi(2t)$) and suppose $\int_0 \varphi(t)dt = \infty$. Then, for every nonnegative subharmonic function $v \not\equiv 0$,*

$$\int_D v(x)\varphi(g(x))\frac{g(x)^2}{\delta(x)^2}dx = \infty.$$

Remark. The special case when $\varphi(t) = t^{-1}$ is particularly interesting. Since one g cancels, we have

$$\int_D v(x)\frac{g(x)}{\delta(x)^2}dx = \infty.$$

COROLLARY 9.5.2. *Let D be a k-Lipschitz domain and let β be as in (9.2). Then*

$$\int_D v(x)\delta(x)^{\beta-2}dx = \infty.$$

One might think that the opposite inequality

$$|\nabla g(x)| \ge A\frac{g(x)}{\delta(x)}$$

holds. This is unfortunately false. Consider a ring domain. For some $t > 0$ the level surface $\{x : g(x) = t\}$ touches itself at some point. Near this point the level surface cannot be written as a graph and $|\nabla g|$ must vanish at the point. Thus the above inequality does not hold.

However, this inequality still holds in a certain sense for Lipschitz domains. Let $\{Q_j\}$ be a Whitney decomposition of D. We may assume that x_0 lies inside Q_0. We shall omit this cube Q_0 in the succeeding argument.

LEMMA 9.5.1. *There exists a positive constant ε_1 such that*

$$|\{x \in Q_j : |\nabla g(x)| \geq \varepsilon_1 \frac{g(x)}{\delta(x)}\}| \approx |Q_j|$$

for every Whitney cube Q_j.

We shall postpone the proof of Lemma 9.5.1. We also need the following weak Harnack inequality for superharmonic functions.

LEMMA 9.5.2. *Let Q be a cube in $\mathbb{R}^n$. For $A_1 > 1$ we let Q^* be the cube with the same center as Q but expanded A_1 times. Suppose E is a measurable subset of Q. If u is a nonnegative superharmonic function on Q^*, then*

$$\frac{1}{|Q|} \int_Q u dx \leq \frac{A_2}{|E|} \int_E u dx,$$

where A_2 depends only on the dimension and A_1.

PROOF. We replace u by the reduced function $\hat{R}_u^Q$ without affecting the integrals. Then we may assume that u is a Green potential in Q^*. Write u as $\int G^*(\cdot, y) d\mu(y)$ with measure μ on the closure of Q, where $G^*(x, y)$ is the Green function for Q^*. Then it is easy to see that

$$Ar^{2-n}|E| \leq \int_E G^*(x, y) dx \leq \int_Q G^*(x, y) dx \leq Ar^2$$

uniformly for y in the closure of Q, where r is the diameter of Q and A depends only on the dimension and A_1. Hence Fubini's theorem yields

$$\frac{1}{|Q|} \int_Q u dx \leq \frac{Ar^2}{|Q|} \mu(\overline{Q}) = Ar^{2-n} \mu(\overline{Q}) \leq \frac{A}{|E|} \int_E u dx.$$

The lemma follows. □

THEOREM 9.5.2. *Let D be a Lipschitz domain domain. Let φ be a doubling function and suppose $\int_0 \varphi(t) dt < \infty$. Then for $r > 0$*

$$\int_{D \setminus B(x_0, r)} u(x) \varphi(g(x)) \frac{g(x)^2}{\delta(x)^2} dx < \infty.$$

PROOF. Since φ is doubling, it follows from the Harnack inequality that

$$\varphi(g(x)) g(x)^2 \delta(x)^{-2} \approx \varphi(g_j) g_j^2 r_j^{-2} \quad \text{for } x \in Q_j,$$

where r_j is the diameter of Q_j and g_j is the value of g at the center of Q_j. Let $E_j = \{x \in Q_j : |\nabla g(x)| \geq \varepsilon_1 g(x)/\delta(x)\}$ and apply Lemma 9.5.2 with $Q = Q_j$ and $E = E_j$. We have from Lemma 9.5.1

$$\int_{Q_j} u\varphi(g) g^2 \delta^{-2} dx \leq A\varphi(g_j) g_j^2 r_j^{-2} \int_{Q_j} u dx \leq A\varphi(g_j) g_j^2 r_j^{-2} \int_{E_j} u dx$$

$$\leq A \int_{E_j} u\varphi(g_j) |\nabla g|^2 dx \leq A \int_{Q_j} u\varphi(g) |\nabla g|^2 dx.$$

Summing over $\{Q_j\}$, we obtain from Theorem 9.5.1 that

$$\int_{D \setminus Q_0} u\varphi(g) g^2 \delta^{-2} dx \leq A \int_{D \setminus Q_0} u\varphi(g) |\nabla g|^2 dx \leq Au(x_0).$$

If necessary, taking the balayage, we may assume that $u(x_0) < \infty$ without changing u in $D \setminus B(x_0, r)$. On any compact subset of $D \setminus \{x_0\}$, $u\varphi(g)g^2\delta^{-2}$ is integrable. Hence the theorem follows. $\square$

Remark. Let $\varphi(t) = t^{\varepsilon-1}$ for a positive constant ε. Then we see

$$\int_D u(x)\frac{g(x)^{1+\varepsilon}}{\delta(x)^2}dx < \infty.$$

Hence we obtain from the lower estimate (9.2.1) that

$$\int_D u(x)\delta(x)^{\varepsilon+\alpha-2}dx < \infty.$$

This shows the weighted integrability of nonnegative superharmonic functions.

In a similar fashion we can show the following L^p-integrability. The proof is omitted. For details we refer to [13].

THEOREM 9.5.3. *Let D be a k-Lipschitz domain and let α and β be as in (9.2.1) and (9.2).*

(i) *If u is a nonnegative superharmonic function on D, then*

$$\int_D u(x)^p dx < \infty$$

for $0 < p < \min\{n/(n+\alpha-2), 1/(\alpha-1)\}$.
(ii) *Let $0 < p \le 1$. If v is a nonnegative nonzero subharmonic function on D, then*

$$\int_D \frac{v(x)^p}{\delta(x)^{n-np+(2-\beta)p}}dx = \infty.$$

We remark that $0 < \beta < 1 < \alpha$ and

$$\min\{n/(n+\alpha-2), 1/(\alpha-1)\} = \begin{cases} n/(n+\alpha-2) & \text{if } 1 < \alpha \le 2, \\ 1/(\alpha-1) & \text{if } \alpha > 2. \end{cases}$$

Remark. In general the calculation of the maximal order $\alpha = \alpha_n(\psi) = \alpha_n(\arctan\frac{1}{k})$ is not so easy. Generally, it requires the Gegenbauer functions. However, for the integer value of the maximal order $\alpha_n(\psi)$ we need only the Gegenbauer polynomials. In particular,

$$\alpha_n(\psi) = 2 \iff \cos\psi = \frac{1}{\sqrt{n}} \iff k = \cot\psi = \frac{1}{\sqrt{n-1}}.$$

This observation is due to F.-Y. Maeda.

Thus we obtain the extension of Masumoto's result.

COROLLARY 9.5.3. *Let D be a k-Lipschitz domain in $\mathbb{R}^n$. If $k < 1/\sqrt{n-1}$, then $u \in L^1(D)$.*

9.6. Lower estimate of the gradient of the Green function. Let us first we observe that a Lipschitz domain is uniformly Δ-regular ([16, Definition 2]). Namely, there are ε_2, $0 < \varepsilon_2 < 1$, and $r_0 > 0$ such that for all $\overline{x} \in \partial D$ and all r, $0 < r < 2r_0$,

$$\omega(\cdot, D \cap \partial B(\overline{x}, r), D \cap B(\overline{x}, r)) \leq 1 - \varepsilon_2 \quad \text{on } D \cap B\left(\overline{x}, \frac{r}{2}\right),$$

where $\omega(\cdot, D \cap \partial B(\overline{x}, r), D \cap B(\overline{x}, r))$ is the harmonic measure of $D \cap \partial B(\overline{x}, r)$ in $D \cap B(\overline{x}, r)$. By an elementary calculation we have from this inequality

$$(9.6.1) \qquad \omega(\cdot, D \cap \partial B(\overline{x}, r), D \cap B(\overline{x}, r)) \leq A\rho^\kappa \quad \text{on } D \cap B(\overline{x}, \rho r)$$

for $0 < \rho < 1$ with $\kappa > 0$ depending only on D.

On the other hand it is known that the boundary Harnack principle holds for a Lipschitz domain (see e.g. [38, Lemma 4.10]). Take $x \in D$ near the boundary. Let $\overline{x} \in \partial D$ such that $|x - \overline{x}| = \delta(x)$. We may assume that $x_0 \in D \setminus B(\overline{x}, 2\delta(x))$. Observe that the functions g and $g(x)\omega(\cdot, D \cap \partial B(\overline{x}, 2\delta(x)), D \cap B(\overline{x}, 2\delta(x)))$ are both positive and harmonic in $D \cap B(\overline{x}, 2\delta(x))$ and vanish on $\partial D \cap B(\overline{x}, 2\delta(x))$. Hence the boundary Harnack principle yields

$$g \approx g(x)\omega(\cdot, D \cap \partial B(\overline{x}, 2\delta(x)), D \cap B(\overline{x}, 2\delta(x))) \quad \text{on } D \cap B(\overline{x}, \delta(x)).$$

In view of (9.6.1) we have

$$g \leq \frac{1}{2}g(x) \quad \text{for } D \cap B(\overline{x}, \rho\delta(x))$$

with sufficiently small $\rho > 0$. In particular, there is $x^* \in B(x, (1 - \rho/2)\delta(x)) \subset D$ such that $g(x^*) \leq \frac{1}{2}g(x)$. Thus we have

LEMMA 9.6.1. *There is a constant c_1, $0 < c_1 < 1$, such that if $x \in D$ and $\delta(x) < r_0$, then there is a point $x^* \in B(x, c_1\delta(x))$ with $g(x^*) \leq \frac{1}{2}g(x)$.*

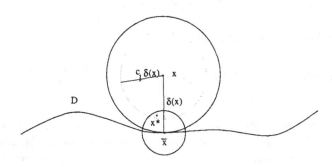

FIGURE 9.3

We are greatful to Professor W. Hansen for drawing our attention to the following general lemma of A. Ancona. This lemma is also used in [14] and simplifies the argument in [13].

LEMMA 9.6.2. *Let $0 < c_2 < c_3 < 1$. Then there is a constant ε, $0 < \varepsilon < 1$, depending only on c_2 and c_3 with the following property: Let h be a positive harmonic function on $B(0,1)$ with $h(0) = 1$. If there is $x \in B(0, c_3)$ such that $h(x) \leq \frac{1}{2}$, then there is $x' \in B(0, c_2)$ such that $h(x') \leq 1 - \varepsilon$.*

PROOF. We prove the lemma by contradiction. Suppose there is no ε with the required property. Then there are positive harmonic functions h_j on $B(0,1)$ such that $h_j(0) = 1$, $\inf_{B(0,c_3)} h_j \leq \frac{1}{2}$ and $\lim_{j\to\infty} \left(\inf_{B(0,c_2)} h_j \right) = 1$. Taking a locally convergent subsequence, we obtain a positive harmonic function h on $B(0,1)$ with $h(0) = 1 = \inf_{B(0,c_2)} h$ and $\inf_{B(0,c_3)} h \leq \frac{1}{2}$. By the minimum principle $h \equiv 1$ on $B(0, c_2)$ and hence on $B(0, 1)$ by analyticity. This is a contradiction. $\square$

By translation and dilation we see from Lemma 9.6.1 and Lemma 9.6.2 that for any c, $0 < c < 1$, there is ε, $0 < \varepsilon < 1$, such that if $x \in D$ and $\delta(x) < r_0$, there there is $x' \in B(x, c\delta(x))$ with $g(x') < (1 - \varepsilon)g(x)$. Let us recall that $\{Q_j\}$ is a Whitney decomposition of D and that x_j and r_j are the center and the diameter of Q_j, respectively. The Harnack inequality says that the value of g is close $g(x_j)$ near x_j and hence

LEMMA 9.6.3. *There are small positive constants c_4 and ε_3 such that if a Whitney cube Q_j is sufficiently small, then*

$$g(x'_j) \leq (1 - \varepsilon_3) \inf_{B(x_j, c_4 r_j)} g$$

for some $x'_j \in Q_j$.

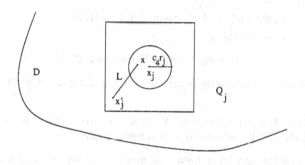

FIGURE 9.4

PROOF OF LEMMA 9.5.1. Observe that $|\nabla g| \leq Ag_j/r_j$ on Q_j, where we recall that r_j is the diameter of Q_j and g_j is the value of g at the center of Q_j. Let $x'_j \in Q_j$ be as in Lemma 9.6.3. For $x \in B(x_j, c_4 r_j)$ we let L be the line segment connecting x and x'_j. Observe that $L \subset Q_j$ and $\ell(L) \leq Ar_j$ uniformly for $x \in Q_j$, where $\ell(L)$ is the length of L. For $\varepsilon > 0$ we have

$$\varepsilon_3 g_j \leq A\varepsilon_3 g(x) \leq A(g(x) - g(x')) \leq A \int_L |\nabla g| ds$$

$$\leq A \int_{\{y \in L : |\nabla g(y)| \leq \varepsilon g_j/r_j\}} \varepsilon \frac{g_j}{r_j} ds + A \int_{\{y \in L : |\nabla g(y)| \geq \varepsilon g_j/r_j\}} A \frac{g_j}{r_j} ds$$

$$\leq A\ell(L)\varepsilon \frac{g_j}{r_j} + A\frac{g_j}{r_j}\ell(\{y \in L : |\nabla g(y)| \geq \varepsilon \frac{g_j}{r_j}\})$$

$$\leq A_3 \varepsilon g_j + A\frac{g_j}{r_j}\ell(\{y \in L : |\nabla g(y)| \geq \varepsilon \frac{g_j}{r_j}\})).$$

Letting $\varepsilon > 0$ so small that $A_3\varepsilon \leq \varepsilon_3/2$, we obtain

$$\frac{\varepsilon_3}{2}g_j \leq A\frac{g_j}{r_j}\ell(\{y \in L : |\nabla g(y)| \geq \varepsilon\frac{g_j}{r_j}\}),$$

whence

$$\ell(\{y \in L : |\nabla g(y)| \geq \varepsilon\frac{g_j}{r_j}\}) \geq Ar_j.$$

This inequality holds for any $x \in B(x_j, c_4r_j)$. Therefore, Fubini's theorem asserts that

$$|\{y \in Q_j : |\nabla g(y)| \geq \varepsilon_3\frac{g_j}{r_j}\}| \geq A|Q_j|.$$

Since $g_j/r_j \approx g(x)/\delta(x)$ for $x \in Q_j$, the lemma follows. □

10. Appendix: Choquet's capacitability theorem

10.1. Analytic sets are capacitable. In this section we give a proof of Choquet's capacitability theorem. This is taken from Bliedtner-Hansen [20, I.3]. Throughout this section we let X be a locally compact Hausdorff space with a countable base. A set function c on X is called a Choquet capacity if the following properties hold:

(i) $0 \leq c(A) \leq +\infty$ for any $A \subset X$.
(ii) If $A \subset B \subset X$, then $c(A) \leq c(B)$.
(iii) If $A_j \uparrow A$, then $c(A_j) \uparrow c(A)$.
(iv) If K_j is compact and $K_j \downarrow K$. then $c(K_j) \downarrow c(K)$.

We say that A is *c-capacitable* if

$$c(A) = \sup\{c(K) : K \text{ compact} \subset A\}.$$

Moreover, A is called *universally capacitable* if A is c-capacitable for any capacity c on X.

Exercise 10.1.1. If μ is a measure on X, then the outer measure μ^* of μ is a capacity. A set is μ^*-capacitable if and only if it is μ-measurable.

Let c be a capacity and let α be a real number. Then the family $\mathcal{C} = \{A \subset X : c(A) > \alpha\}$ enjoys the following properties:

(i) If $A \in \mathcal{C}$ and $A \subset A' \subset X$, then $A' \in \mathcal{C}$.
(ii) If A_j is increasing and $\bigcup_{j=1}^{\infty} A_j \in \mathcal{C}$, then $A_j \in \mathcal{C}$ for some j.

In fact, (i) and (ii) follows from (ii) and (iii) of the properties of capacity.

Remark 10.1.1. In general, a family $\mathcal{C}$ of subsets of X with (i) and (ii) is called a *capacitance*.

Let us define analytic sets. For a locally compact space Z we let, in general,

$$K_{\sigma\delta}(Z) = \{\bigcap_{i=1}^{\infty} \bigcup_{j=1}^{\infty} K_{ij} : K_{ij} \text{ compact} \subset Z\}.$$

We say that $A \subset X$ is *analytic* if there are a compact space Y with a countable base and $B \in K_{\sigma\delta}(X \times Y)$ such that $A = \pi_X(B)$, where π_X is the projection of $X \times Y$ on X.

THEOREM 10.1.1. *(G. Choquet) Every analytic subset of X is universally capacitable.*

PROOF. Let A be an analytic subset of X and let c be a capacity on X. Take an arbitrary number $\alpha < c(A)$. It is sufficient to find a compact subset K of A such that $c(K) \geq \alpha$.

Let $\mathcal{C} = \{E \subset X : c(E) > \alpha\}$. Obviously, $A \in \mathcal{C}$. By definition there are a compact space Y with a countable base and compact subsets K_{ij} of $X \times Y$ such that

$$\pi_X(B) = A \quad \text{with } B = \bigcap_{i=1}^{\infty} \bigcup_{j=1}^{\infty} K_{ij}.$$

We may assume that K_{ij} is increasing for each fixed i. By definition $B \subset \bigcup_{j=1}^{\infty} K_{1j}$ and hence $B = \bigcup_{j=1}^{\infty} B \cap K_{1j}$, which implies that $A = \pi_X(B) = \bigcup_{j=1}^{\infty} \pi_X(B \cap K_{1j})$. Hence, by (ii) for $\mathcal{C}$, there is m_1 such that

$$A_1 = \pi_X(B \cap K_{1m_1}) \in \mathcal{C}.$$

Obviously, A_1 is a relatively compact subset of A. By definition $B \subset \bigcup_{j=1}^{\infty} K_{2j}$ and hence $B \cap K_{1m_1} = \bigcup_{j=1}^{\infty} B \cap K_{1m_1} \cap K_{2j}$, which implies that $A_1 = \pi_X(B \cap K_{1m_1}) = \bigcup_{j=1}^{\infty} \pi_X(B \cap K_{1m_1} \cap K_{2j})$. Hence, again by (ii) for $\mathcal{C}$, there is m_2 such that

$$A_2 = \pi_X(B \cap K_{1m_1} \cap K_{2m_2}) \in \mathcal{C}.$$

Repeat this procedure. We can always choose A_j in such a way that

$$A_j = \pi_X(B \cap K_{1m_1} \cap \cdots \cap K_{jm_j}) \in \mathcal{C}.$$

Let $K_j = \bigcap_{i=1}^{j} K_{im_i}$. Then $\overline{A_j} \subset \pi_X(K_j)$. Since K_j is a decreasing sequence of compact subsets of $X \times Y$, it follows that $\bigcap_{j=1}^{\infty} \pi_X(K_j) = \pi_X(\bigcap_{j=1}^{\infty} K_j)$ (see Exercise 10.1.2 below), so that

$$K := \bigcap_{j=1}^{\infty} \overline{A_j} \subset \bigcap_{j=1}^{\infty} \pi_X(K_j) = \pi_X(\bigcap_{j=1}^{\infty} K_j) \subset \pi_X(B) = A.$$

We observe that K is a compact subset of A and

$$c(K) = \lim_{j \to \infty} c(\overline{A_j}) \geq \lim_{j \to \infty} c(A_j) \geq \alpha$$

by the property (iv) of capacity. The proof is complete. $\square$

Exercise 10.1.2. Let f be a continuous map on a locally compact space X with countable base and let $K_j \subset X$ be a decreasing sequence of compact sets. Prove $f(\bigcap_{j=1}^{\infty} K_j) = \bigcap_{j=1}^{\infty} f(K_j)$.

10.2. Borel sets are analytic. In this section we shall show that Borel sets are analytic. For this purpose we prove that countable union and countable intersection of analytic sets are analytic.

THEOREM 10.2.1. *Let A_j be analytic sets in X. Then $\bigcup_{j=1}^{\infty} A_j$ and $\bigcap_{j=1}^{\infty} A_j$ are analytic.*

PROOF. By definition there are compact spaces Y_j with countable base and $B_j \in K_{\sigma\delta}(X \times Y_j)$ such that $\pi_X(B_j) = A_j$. Without loss of generality we may assume that $\{Y_j\}$ is disjoint. Let Y be the one point compactification of $\bigcup_{j=1}^{\infty} Y_j$. Then Y is a compact space with countable base. Since $B_j \in K_{\sigma\delta}(X \times Y_j)$, we can write $B_j = \bigcap_{i=1}^{\infty} E_{ji}$ with $E_{ji} \in K_{\sigma}(X \times Y_j)$. Since $E_{ji} \subset X \times Y_j$ and $\{Y_j\}$ is disjoint, it follows that

$$\bigcup_{j=1}^{\infty} B_j = \bigcup_{j=1}^{\infty} \left(\bigcap_{i=1}^{\infty} E_{ji} \right) = \bigcap_{i=1}^{\infty} \left(\bigcup_{j=1}^{\infty} E_{ji} \right)$$

(see Exercise 10.2.1 below), so that $\bigcup_{j=1}^{\infty} B_j \in K_{\sigma\delta}(X \times Y)$. Hence $\bigcup_{j=1}^{\infty} A_j = \bigcup_{j=1}^{\infty} \pi_X(B_j) = \pi_X(\bigcup_{j=1}^{\infty} B_j)$ is analytic.

In order to show that $\bigcap_{j=1}^{\infty} A_j$ is analytic, we consider the product space $Z = \prod_{j=1}^{\infty} Y_j$. This is a compact space. By π_j we denote the projection of $X \times Z$ on $X \times Y_j$. Observe that if K is a compact subset of $X \times Y_j$, then $\pi_j^{-1}(K)$ is a compact subset of $X \times Z$ (see Exercise 10.2.2 below). Hence, $B_j' = \pi_j^{-1}(B_j) \in K_{\sigma\delta}(X \times Z)$, so that $B' = \bigcap_{j=1}^{\infty} B_j' \in K_{\sigma\delta}(X \times Z)$. Hence it suffices to show that

$$\pi_X(B') = \bigcap_{j=1}^{\infty} A_j.$$

Since $\pi_X(B') \subset \pi_X(B_j') = A_j$ for each j, we have $\pi_X(B') \subset \bigcap_{j=1}^{\infty} A_j$. Let us prove the reverse inclusion. Let $x \in \bigcap_{j=1}^{\infty} A_j$. Then for each j there is $y_j \in Y_j$ such that $(x, y_j) \in B_j$. Letting $z = (y_1, y_2, \dots) \in Z$, we obtain $\pi_j(x, z) = (x, y_j) \in B_j$, and hence $(x, z) \in \bigcap_{j=1}^{\infty} \pi_j^{-1}(B_j) = \bigcap_{j=1}^{\infty} B_j'$. Hence $x \in \pi_X(B')$. The proof is complete. $\square$

Exercise 10.2.1. Suppose $\{E_{ji}\}_{j=1}^{\infty}$ are disjoint for every i. Then

$$\bigcup_{j=1}^{\infty} \left(\bigcap_{i=1}^{\infty} E_{ji} \right) = \bigcap_{i=1}^{\infty} \left(\bigcup_{j=1}^{\infty} E_{ji} \right).$$

Show that the assumption of disjointness cannot be dropped.

Exercise 10.2.2. Let X and Y be compact Hausdorff spaces and let $f : X \to Y$ be continuous. If K is a compact subset of Y, then $f^{-1}(K)$ is a compact subset of X.

THEOREM 10.2.2. *Every Borel set is analytic.*

PROOF. For $A \subset X$, we denote by $\mathsf{C}A$ the complement $X \setminus A$. Let $\mathcal{M}$ be the family of analytic subsets A of X such that $\mathsf{C}A$ is analytic. By definition $\mathcal{M}$ is closed under complementation. Let us show that $\mathcal{M}$ is closed under the formation of countable unions and intersections. Suppose $A_j \in \mathcal{M}$. Then $\mathsf{C}A_j \in \mathcal{M}$ and Theorem 10.2.1 says $\bigcup_{j=1}^{\infty} A_j$, $\bigcap_{j=1}^{\infty} A_j \bigcup_{j=1}^{\infty} \mathsf{C}A_j$ and $\bigcap_{j=1}^{\infty} \mathsf{C}A_j$ are altogether analytic sets. Hence $\bigcup_{j=1}^{\infty} A_j, \bigcap_{j=1}^{\infty} A_j \in \mathcal{M}$. Thus, $\mathcal{M}$ is a σ-algebra.

Obviously, every compact set is analytic. In fact, if K is compact, then $K \times \{0\} \in K_{\sigma\delta}(K \times \{0\})$ and $K = \pi_X(K \times \{0\})$. Observe that an open subset of X and its complement are written as the countable union of compact sets. Hence, open sets belong to $\mathcal{M}$. Since the family of Borel sets is the smallest σ-algebra including all open sets, it follows that the family of Borel sets is a subfamily of $\mathcal{M}$. This means that every Borel set is analytic. The proof is complete. $\square$

By Theorem 10.1.1 and Theorem 10.2.2 we have

COROLLARY 10.2.1. *Every Borel set is universally capacitable.*

11. Appendix: Minimal fine limit theorem

11.1. Introduction. The purpose of this appendix is to show the minimal fine limit theorem. There is a systematic treatise by Doob [25]. However, we follow Brelot [21] to make the argument shorter. The reader is assumed to be familiar with the Martin boundary theory and the Perron-Wiener-Brelot method. For these accounts we refer to Helms [36]. For convenience, we collect some basic facts of the Martin boundary theory.

Throughout this appendix we let Ω be a domain with Green function $G(x, y)$. Let $x_0 \in \Omega$ and let $K(x, y) = G(x, y)/G(x_0, y)$. The function $K(x, y)$ extends continuously to $\Omega \times \widehat{\Omega}$, where $\widehat{\Omega}$ is the Martin compactification. The extension is denoted also by $K(x, y)$ and is called the Martin kernel for Ω. For simplicity we write $K_y = K(\cdot, y)$. The Martin compactification $\widehat{\Omega}$ is metrizable. The boundary $\Delta = \widehat{\Omega} \setminus \Omega$ is called the Martin boundary. The Martin boundary is separated by the family $\{K(x, \cdot)\}_{x \in \Omega}$. We let $\Delta_1 = \{y \in \Delta : K_y \text{ is a minimal harmonic function}\}$ and let $\Delta_0 = \Delta \setminus \Delta_1$. We say that Δ_1 is the minimal boundary.

For every nonnegative harmonic function h on Ω there exists a unique measure μ_h on Δ_1 such that $h = \int_{\Delta_1} K_y d\mu_h(y)$. This is the Martin representation theorem. In view of the Riesz decomposition theorem, for every nonnegative superharmonic function u on Ω, there exists a unique measure μ_u on $\Omega \cup \Delta_1$ such that $u = \int_{\Omega \cup \Delta_1} K_y d\mu_u(y)$. This is the Riesz-Martin representation theorem.

11.2. Balayage (Reduced function).

Let us begin by recalling balayage or the theory of reduced functions.

DEFINITION. Let u be a nonnegative superharmonic function on Ω and $E \subset \Omega$. We let

$$R_u^E = \inf\{v : v \geq 0, \text{ superharmonic}, v \geq u \text{ on } E\}.$$

The lower regularization of R_u^E is denoted by $\widehat{R}_u^E$ and is called the regularized reduced function or the balayage of u to E.

PROPOSITION 11.2.1. *We enlist some properties of regularized reduced functions:*

(i) $\widehat{R}_u^E$ *is superharmonic.*

(ii) $\widehat{R}_u^E \leq R_u^E$ *and the equality holds outside a polar set.*

(iii) $\widehat{R}_u^E = R_u^E$ *on* $\Omega \setminus \overline{E}$ *and is harmonic there.*

(iv) $\widehat{R}_u^E$ *is the smallest nonnegative superharmonic function which majorizes* u *q.e. on* E.

(v) *If* $E_j \uparrow E$, *then* $\widehat{R}_u^{E_j} \uparrow \widehat{R}_u^E$.

(vi) *If* u *is continuous, then* $\widehat{R}_u^E = \inf_{\substack{U \supset E \\ U \text{ open}}} \widehat{R}_u^U$.

(vii) *Let* E *be a compact subset and let* $H_u^{\Omega \setminus E}$ *be the PWB solution on* $\Omega \setminus E$ *with the boundary function* u. *Moreover, let* $\omega_{\Omega \setminus E}^x$ *be the harmonic measure at* x *of* $\Omega \setminus E$. *Then*

$$\widehat{R}_u^E(x) = H_u^{\Omega \setminus E}(x) = \int u \, d\omega_{\Omega \setminus E}^x.$$

DEFINITION. If $u = G\mu$ is a Green potential, then $\widehat{R}_{G\mu}^E$ is a Green potential. By the Riesz theorem there is a measure b_μ^E such that

$$\widehat{R}_{G\mu}^E = Gb_\mu^E.$$

Note that b_μ^E is supported on $\overline{E}$, because $\widehat{R}_{G\mu}^E$ is harmonic on $\Omega \setminus \overline{E}$. We call b_μ^E the balayage measure. If μ is the unit measure $\mathcal{E}_x$ at x, then we write $b_{\mathcal{E}_x}^E$ for b_μ^E.

We write $G_x = G\mathcal{E}_x$. Then $\widehat{R}_{G_x}^E = Gb_{\mathcal{E}_x}^E$. We have the following reciprocity relation.

THEOREM 11.2.1. *Let* $E \subset \Omega$ *and let* $x, y \in \Omega$. *Then*

$$\widehat{R}_{G_x}^E(y) = \widehat{R}_{G_y}^E(x).$$

PROOF. We may assume that E is compact. First suppose that $x, y \notin E$. Let $\mathcal{G}$ be the Green function for $\Omega \setminus E$. Observe that $G_x - \hat{R}_{G_x}^E$ is nonnegative and harmonic on $\Omega \setminus (E \cup \{x\})$ and has a singularity at x. Thus, this is the Green function for $\Omega \setminus E$ with pole at x. Hence $\mathcal{G}_x = G_x - \hat{R}_{G_x}^E$, or

$$\mathcal{G}(x, y) = G(x, y) - \hat{R}_{G_x}^E(y).$$

Changing the roles of x and y,

$$\mathcal{G}(y, x) = G(y, x) - \hat{R}_{G_y}^E(x).$$

By the symmetry of $\mathcal{G}$ and G we have $\hat{R}_{G_x}^E(y) = \hat{R}_{G_y}^E(x)$. If x or y lies in E, then we remove a small ball with center at x or y, apply the above argument and shrink the ball. $\square$

Let $x \in \Omega$ and let E be a compact subset of $\Omega \setminus \{x\}$. By using the harmonic measure we can write the balayage as

$$\hat{R}_{G_y}^E(x) = \int G_y d\omega_{\Omega \setminus E}^x = \int G(y, z) d\omega_{\Omega \setminus E}^x(z).$$

On the other hand, by using the balayage measure we can write the balayage as

$$\hat{R}_{G_x}^E(y) = G b_{\mathcal{E}_x}^E(y) = \int G(y, z) d b_{\mathcal{E}_x}^E(z).$$

By the reciprocity (Theorem 11.2.1)

$$\int G(y, z) d\omega_{\Omega \setminus E}^x(z) = \int G(y, z) d b_{\mathcal{E}_x}^E(z).$$

Since the family $\{G(y, \cdot)\}_{y \in \Omega}$ is ample, $\omega_{\Omega \setminus E}^x = b_{\mathcal{E}_x}^E$. Thus we have from Proposition 11.2.1 (vii)

$$\hat{R}_u^E(x) = \int u d\omega_{\Omega \setminus E}^x = \int u d b_{\mathcal{E}_x}^E.$$

The left and the right hand sides have meaning for every E. Hence the approximation yields the following theorem.

THEOREM 11.2.2. *Let $E \subset \Omega$ and $x \in \Omega$. If u is a nonnegative superharmonic function on Ω, then*

$$\hat{R}_u^E(x) = \int u d b_{\mathcal{E}_x}^E.$$

If a nonnegative superharmonic function is represented by the Riesz-Martin theorem, then its balayage is obtained by the balayage of the Martin kernel.

THEOREM 11.2.3. *Let u be a nonnegative superharmonic function and suppose it has the Riesz-Martin representation*

$$u = \int_{\Omega \cup \Delta_1} K_y d\mu_u(y).$$

Then

$$\hat{R}_u^E = \int_{\Omega \cup \Delta_1} \hat{R}_{K_y}^E d\mu_u(y).$$

PROOF. By the balayage measure (Theorem 11.2.2)

$$\hat{R}_u^E(x) = \int u \, db_{\mathcal{E}_x}^E = \int (\int K_y d\mu_u(y)) db_{\mathcal{E}_x}^E = \int (\int K_y db_{\mathcal{E}_x}^E) d\mu_u(y)$$
$$= \int \hat{R}_{K_y}^E d\mu_u(y).$$

$\square$

11.3. Minimal thinness. In this subsection we define minimal thinness and state some properties.

DEFINITION. Let $y \in \Delta_1$ and $E \subset \Omega$. We say that E is minimally thin at y if $\hat{R}_{K_y}^E \neq K_y$.

LEMMA 11.3.1. *A set E is minimally thin at $y \in \Delta_1$ if and only if $\hat{R}_{K_y}^E$ is a Green potential.*

PROOF. Assume that E is minimally thin at y. By the Riesz decomposition we have

$$\hat{R}_{K_y}^E = v + h,$$

where v is a Green potential and h is the greatest harmonic minorant. Since $h \leq K_y$, the minimality of K_y implies $h = cK_y$ with $0 \leq c < 1$. Suppose $0 < c < 1$ and we shall derive a contradiction. By definition

$$\hat{R}_{\hat{R}_u^E}^E - \hat{R}_u^E$$

in general. Hence

$$\hat{R}_{v+cK_y}^E = v + cK_y.$$

On the other hand, the linearity yields

$$\hat{R}_{v+cK_y}^E = \hat{R}_v^E + c\hat{R}_{K_y}^E = \hat{R}_v^E + c(v + cK_y) \leq v + c(v + cK_y).$$

Therefore

$$cK_y \leq c(v + cK_y)$$

and so K_y is dominated by the Green potential $v/(1-c)$. This is a contradiction. The opposite implication is obvious. $\square$

LEMMA 11.3.2. *Let $\mathcal{V} \neq \emptyset$ be a neighborhood of $y \in \Delta_1$ in $\hat{\Omega}$. Then*

(i) $\Omega \setminus \mathcal{V}$ *is minimally thin at y.*

(ii) $\mathcal{V} \cap \Omega$ *is not minimally thin at y.*

PROOF. (i) Suppose the contrary: $\hat{R}_{K_y}^{\Omega \setminus \mathcal{V}} = K_y$. Let Ω_j be relatively compact open sets such that $\Omega_j \uparrow \Omega$. Then

$$\hat{R}_{K_y}^{\Omega_j \setminus \mathcal{V}} \uparrow K_y.$$

Since Ω_j is compact, the left hand side is a Green potential. So we find a measure μ_j such that

$$K\mu_j = \hat{R}_{K_y}^{\Omega_j \setminus \mathcal{V}}.$$

Evaluating at x_0, we have

$$\|\mu_j\| = K\mu_j(x_0) = \hat{R}_{K_y}^{\Omega_j \setminus \mathcal{V}}(x_0) \leq K_y(x_0) = 1.$$

Thus, if necessary taking a subsequence, we may assume that μ_j converges vaguely to a measure μ on $\hat{\Omega}$. We see that $K\mu_j \to K\mu$ on Ω, and hence $K\mu = K_y$. Since μ_j is supported on the boundary of $\Omega_j \setminus \mathcal{V}$, it follows that μ is supported on $\Delta \setminus \mathcal{V}$. Hence $\mu(\mathcal{V}) = 0$ so that $\mu \neq \mathcal{E}_y$. Thus we have two different expressions for K_y,

$$K_y = K\mathcal{E}_y = K\mu.$$

This contradicts the uniqueness of the representation.

(ii) By (i) and Lemma 11.3.1 $\hat{R}_{K_y}^{\Omega \setminus \mathcal{V}}$ is a Green potential, and hence its greatest harmonic minorant is 0. Observe

$$K_y = \hat{R}_{K_y}^{\Omega} \leq \hat{R}_{K_y}^{\mathcal{V} \cap \Omega} + \hat{R}_{K_y}^{\Omega \setminus \mathcal{V}}.$$

Take the greatest harmonic minorants of the both sides. Observe that K_y is majorized by the greatest harmonic minorant of $\hat{R}_{K_y}^{\mathcal{V} \cap \Omega}$ and so by $\hat{R}_{K_y}^{\mathcal{V} \cap \Omega}$ itself. Hence $K_y = \hat{R}_{K_y}^{\mathcal{V} \cap \Omega}$. Thus $\mathcal{V} \cap \Omega$ is not minimally thin at y. $\square$

We can separate a point $y \in \Delta_1$ from the remaining boundary points by an open set in $\hat{\Omega}$, whose portion in Ω is minimally thin at y.

LEMMA 11.3.3. *Let $y \in \Delta_1$. There is an open set $\mathcal{V}$ in $\hat{\Omega}$ such that $\mathcal{V} \cap \Omega$ is minimally thin at y and $\mathcal{V} \supset \Delta \setminus \{y\}$.*

The following lemma is elementary.

LEMMA 11.3.4. *Let u be a nonnegative superharmonic function on Ω. Suppose $\Omega_j \uparrow \Omega$. If $\hat{R}_u^{\Omega \setminus \Omega_j}$ is a Green potential for some j, then $\hat{R}_u^{\Omega \setminus \Omega_j} \to 0$.*

PROOF. Let $h = \lim_{j \to \infty} \hat{R}_u^{\Omega \setminus \Omega_j}$. Observe that h is a nonnegative harmonic function. This is a harmonic minorant of $\hat{R}_u^{\Omega \setminus \Omega_j}$, which is a Green potential. Hence $h = 0$. $\square$

PROOF OF LEMMA 11.3.3. Take an increasing sequence of open sets $\mathcal{V}_j$ such that $\bigcup \mathcal{V}_j = \hat{\Omega} \setminus \{y\}$ and $y \notin \overline{\mathcal{V}_j}$, where $\overline{\mathcal{V}_j}$ is the closure in $\hat{\Omega}$. By Lemma 11.3.2. $\mathcal{V}_j \cap \Omega$ is minimally thin at y. Hence $\hat{R}_{K_y}^{\mathcal{V}_j \cap \Omega}$ is a Green potential. By Lemma 11.3.4 we can find an open set $\mathcal{W}_j$ in $\hat{\Omega}$ such that $\Delta \subset \mathcal{W}_j$ and $\hat{R}_{K_y}^{\mathcal{V}_j \cap \mathcal{W}_j \cap \Omega}(x_0) < 2^{-j}\varepsilon$ with $0 < \varepsilon < 1$. Let $\mathcal{V} = \bigcup_j (\mathcal{V}_j \cap \mathcal{W}_j)$. Then

$$\hat{R}_{K_y}^{\mathcal{V} \cap \Omega}(x_0) \leq \sum \hat{R}_{K_y}^{\mathcal{V}_j \cap \mathcal{W}_j \cap \Omega}(x_0) < \varepsilon < 1.$$

Thus $\mathcal{V} \cap \Omega$ is minimally thin at y. Obviously $\mathcal{V} \supset \Delta \setminus \{y\}$. $\square$

DEFINITION. For $E \subset \Omega$ we put

$$\mathcal{E}_E = \{y \in \Delta_1 : E \text{ is minimally thin at } y\},$$

$$\overline{E}^{\mathrm{mf}} = E \cup \{y \in \Delta_1 : E \text{ is not minimally thin at } y\}.$$

We call $\overline{E}^{\mathrm{mf}}$ the minimal fine closure of E. Observe that $\mathcal{E}_E = \Delta_1 \setminus \overline{E}^{\mathrm{mf}}$.

THEOREM 11.3.1. *Let u be a positive harmonic function on Ω with the Martin representing measure μ_u. Then*

$$\hat{R}_u^E = u \iff \mu_u(\Delta_1 \setminus \overline{E}^{mf}) = 0.$$

PROOF. By Theorem 11.2.3

$$\hat{R}_u^E(x) = \int_{\Delta_1} \hat{R}_{K_y}^E(x)d\mu_u(y) \geq \int_{\Delta_1 \setminus \mathcal{E}_E} \hat{R}_{K_y}^E(x)d\mu_u(y) = \int_{\Delta_1 \setminus \mathcal{E}_E} K_y(x)d\mu_u(y).$$

Suppose $\mu_u(\mathcal{E}_E) = 0$. Then the above inequality becomes the equality. Hence

$$\hat{R}_u^E(x) = \int_{\Delta_1 \setminus \mathcal{E}_E} K_y(x)d\mu_u(y) = \int_{\Delta_1} K_y(x)d\mu_u(y) = u(x).$$

Thus $\hat{R}_u^E = u$.

Conversely, suppose $\hat{R}_u^E = u$. Then

$$\int_{\Delta_1} K_y d\mu_u(y) = u = \hat{R}_u^E = \int_{\Delta_1} \hat{R}_{K_y}^E d\mu_u(y).$$

Since $\hat{R}_{K_y}^E \leq K_y$, we have $\hat{R}_{K_y}^E = K_y$ μ_u-a.e. For $y \in \mathcal{E}_E$ we have $\hat{R}_{K_y}^E \neq K_y$, so that $\mu_u(\mathcal{E}_E) = 0$. $\square$

DEFINITION. Let v be a real function on Ω and let $y \in \Delta_1$. We define

$$\operatorname{mf\,limsup}_{x \to y} v(x) = \sup\{\alpha : \{z : v(z) > \alpha\} \text{ is not minimally thin at } y\},$$

$$\operatorname{mf\,liminf}_{x \to y} v(x) = \inf\{\alpha : \{z : v(z) < \alpha\} \text{ is not minimally thin at } y\}.$$

If $\operatorname{mf\,limsup}_{x \to y} v(x) = \operatorname{mf\,liminf}_{x \to y} v(x)$, then we write $\operatorname{mf\,lim}_{x \to y} v(x)$ for this value and call it the minimal fine limit of v at y.

Remark.

$$\liminf v \leq \operatorname{mf\,liminf} v \leq \operatorname{mf\,limsup} v \leq \limsup v.$$

Remark. Observe that $\alpha = \operatorname{mf\,limsup}_{x \to y} v(x)$ if and only if for each $\varepsilon > 0$ the following two conditions hold:

  (i) $\{z : v(z) > \alpha + \varepsilon\}$ is minimally thin at y.
  (ii) $\{z : v(z) > \alpha - \varepsilon\}$ is not minimally thin at y.

By using the minimal fine limit we can state the following extended minimum principle. We emphasize that "mf limsup" and not "mf liminf" appears in the theorem. This is quite a difference from the usual Phragmén-Lindelöf type minimum principle.

THEOREM 11.3.2. *Let h be a positive harmonic function on Ω. Let u be superharmonic and suppose $u/h \geq -A$ on Ω with some positive constant A. If*

$$\operatorname{mf\,limsup} \frac{u}{h} \geq 0 \ \mu_h\text{-}a.e. \ on \ \Delta_1,$$

then $u \geq 0$ on Ω.

PROOF. Let $u_1 = \min\{u, 0\}$. It is sufficient to prove $u_1 \equiv 0$. Let u_1' be the greatest harmonic minorant of u_1. Observe

(11.3.1) $$0 \geq u_1' \geq -Ah.$$

We claim $u_1' = 0$. For $\varepsilon > 0$ we let

$$E_\varepsilon = \{x : u_1(x) > -\varepsilon h(x)\}.$$

By definition $E_\varepsilon = \{x : u(x)/h(x) > -\varepsilon\}$. By assumption for μ_h-a.e. $y \in \Delta_1$ E_ε is not minimally thin at y. Hence

$$\mu_h(\Delta_1 \setminus \overline{E_\varepsilon}^{\text{mf}}) = 0.$$

By (11.3.1) the Martin representing measure $\mu_{-u_1'}$ of $-u_1'$ is absolutely continuous with respect to μ_h. Hence

$$\mu_{-u_1'}(\Delta_1 \setminus \overline{E_\varepsilon}^{\mathrm{mf}}) = 0.$$

By Theorem 11.3.1

$$(11.3.2) \qquad\qquad\qquad \hat{R}^{E_\varepsilon}_{-u_1'} = -u_1'.$$

By the Riesz decomposition $u_1 = u_1' + v$ with Green potential v. By definition $u_1 > -\varepsilon h$ on E_ε and hence

$$-u_1' < \varepsilon h + v \quad \text{on } E_\varepsilon.$$

Take the balayage of the both sides. By (11.3.2)

$$-u_1' = \hat{R}^{E_\varepsilon}_{-u_1'} \le \hat{R}^{E_\varepsilon}_{\varepsilon h + v} \le \varepsilon h + v \quad \text{on } \Omega.$$

Since $\varepsilon > 0$ is arbitrary, it follows that $-u_1' \le v$. Since v is a Green potential, it follows that $u_1' \equiv 0$ and hence u_1 is a Green potential. But $u_1 \le 0$, so this potential must be identically equal to 0. Thus $u_1 \equiv 0$. The theorem follows. $\square$

11.4. PWB^h solution. In this subsection we introduce PWB^h solutions and give some properties.

DEFINITION. Let h be a fixed positive harmonic function on Ω. For f on Δ we put

$$\mathcal{U}_{f,h} = \{w \text{ superharmonic}: \liminf \frac{w}{h} \ge f \text{ on } \Delta\},$$

$$\mathcal{L}_{f,h} = \{w \text{ subharmonic}: \limsup \frac{w}{h} \le f \text{ on } \Delta\}.$$

We put

$$\overline{\mathcal{D}}_{f,h} = \inf\{w : w \in \mathcal{U}_{f,h}\},$$

$$\underline{\mathcal{D}}_{f,h} = \sup\{w : w \in \mathcal{L}_{f,h}\}.$$

If $\overline{\mathcal{D}}_{f,h} = \underline{\mathcal{D}}_{f,h}$, then we say f is h-resolutive and write $\overline{\mathcal{D}}_{f,h} = \underline{\mathcal{D}}_{f,h} = \mathcal{D}_{f,h}$. It is known that a Borel measurable function is h-resolutive. We can write

$$\mathcal{D}_{f,h}(x) = \int_\Delta f \, d\nu_h^x,$$

where ν_h^x is called the h-harmonic measure at x.

Remark. Doob [25] uses a different notation; he writes

$$\frac{1}{h}\mathcal{D}_{f,h}(x) = \int_\Delta f \, d\omega_h^x.$$

We call $\omega_h^x = \frac{1}{h}\nu_h^x$ the h-harmonic measure at x.

Let $\mathcal{E} \subset \Delta$ and let h be a positive harmonic function. We write $h_\mathcal{E} = \overline{\mathcal{D}}_{\chi_\mathcal{E},h}$. This is a nonnegative harmonic function on Ω.

LEMMA 11.4.1.

$$h_\mathcal{E} = \inf_{\substack{V \text{ open} \\ V \supset \mathcal{E}}} \hat{R}_h^{V \cap \Omega}.$$

PROOF. By definition $h_\mathcal{E} \leq \widehat{R}_h^{\mathcal{V}\cap\Omega}$ for any neighborhood of $\mathcal{E}$. Thus '$\leq$' follows. Conversely, let u be a nonnegative superharmonic function such that

$$\liminf \frac{u}{h} \geq \chi_\mathcal{E}.$$

By the l.s.c.

$$\mathcal{V} = \{x \in \Omega : \frac{u(x)}{h(x)} > 1 - \varepsilon\} \cup \{x \in \Delta : \liminf_{y \to x} \frac{u(y)}{h(y)} > 1 - \varepsilon\}$$

is an open neighborhood of $\mathcal{E}$ for any $\varepsilon > 0$. Let $0 < \varepsilon < 1$. We have $\dfrac{u}{1 - \varepsilon} \geq h$ on $\mathcal{V} \cap \Omega$. Hence

$$\frac{u}{1 - \varepsilon} \geq \widehat{R}_h^{\mathcal{V}\cap\Omega} \geq \inf_{\substack{\mathcal{V} \text{ open} \\ \mathcal{V} \supset \mathcal{E}}} \widehat{R}_h^{\mathcal{V}\cap\Omega}.$$

Take the infimum with respect to u:

$$\frac{h_\mathcal{E}}{1 - \varepsilon} = \frac{1}{1 - \varepsilon} \overline{\mathcal{D}}_{\chi_\mathcal{E}, h} \geq \inf_{\substack{\mathcal{V} \text{ open} \\ \mathcal{V} \supset \mathcal{E}}} \widehat{R}_h^{\mathcal{V}\cap\Omega}.$$

Let $\varepsilon \to 0$. We obtain

$$h_\mathcal{E} \geq \inf_{\substack{\mathcal{V} \text{ open} \\ \mathcal{V} \supset \mathcal{E}}} \widehat{R}_h^{\mathcal{V}\cap\Omega}.$$

$\square$

LEMMA 11.4.2. Let $y \in \Delta_1$. Then

$$(K_y)_\mathcal{E} = \begin{cases} K_y & \text{if } y \in \mathcal{E}, \\ 0 & \text{if } y \notin \mathcal{E}. \end{cases}$$

PROOF. Suppose $y \in \mathcal{E}$. Then for any neighborhood $\mathcal{V}$ of y, $\mathcal{V} \cap \Omega$ is not minimally thin at y by Lemma 11.3.2 (ii), or in other words

$$\widehat{R}_{K_y}^{\mathcal{V}\cap\Omega} = K_y.$$

By Lemma 11.4.1 $(K_y)_\mathcal{E} = K_y$. For the remaining case it is sufficient to show that

$$(K_y)_{\Delta\setminus\{y\}} = 0.$$

By Lemma 11.3.3 we can find a neighborhood $\mathcal{V}$ of $\Delta \setminus \{y\}$ such that $\mathcal{V}$ is minimally thin at y. By Lemma 11.4.1

$$(K_y)_{\Delta\setminus\{y\}} \leq \widehat{R}_{K_y}^{\mathcal{V}\cap\Omega},$$

and the right hand side is a Green potential by Lemma 11.3.1. Hence $(K_y)_{\Delta\setminus\{y\}} = 0$. $\square$

LEMMA 11.4.3. Let $\mathcal{E}$ be a Borel subset of Δ. Then

$$h_\mathcal{E} = \int_\mathcal{E} K_y d\mu_h(y).$$

PROOF. We may assume that $\mathcal{E}$ is compact. Let $\mathcal{V}$ be a neighborhood of $\mathcal{E}$ in $\hat{\Omega}$. By Theorem 11.2.3

$$\hat{R}_h^{\mathcal{V} \cap \Omega} = \int \hat{R}_{K_y}^{\mathcal{V} \cap \Omega} d\mu_h(y).$$

Apply this to $\mathcal{V} = \mathcal{V}_j$ and let $\mathcal{V}_j \to \mathcal{E}$. By Lemma 11.4.1 and Lemma 11.4.2

$$h_{\mathcal{E}} = \int (K_y)_{\mathcal{E}} d\mu_h(y) = \int_{\mathcal{E}} K_y d\mu_h(y).$$

$\square$

Let us give a relationship between h-harmonic measures and Martin representing measures.

THEOREM 11.4.1. *The h-harmonic measure is given by* $d\nu_h^x = K_y(x)d\mu_h(y).$

PROOF. We have from Lemma 11.4.3

$$\int_{\Delta} \chi_{\mathcal{E}} d\nu_h^x = \mathcal{D}_{\chi_{\mathcal{E}},h}(x) = h_{\mathcal{E}}(x) = \int_{\mathcal{E}} K_y(x)d\mu_h(y)$$

fore every Borel set $\mathcal{E}$. Hence $d\nu_h^x = K_y(x)d\mu_h(y).$ $\square$

The following theorem is the minimum principle for a PWB^h solution.

THEOREM 11.4.2. *Let f be a bounded h-resolutive function on Δ. Let u be super-harmonic and suppose $u/h \geq -A$ on Ω with some positive constant A. If*

$$\text{mf} \limsup \frac{u}{h} \geq f \ \mu_h\text{-a.e. on } \Delta_1,$$

then $u \geq \mathcal{D}_{f,h}$ on Ω.

PROOF. Note

$$\mathcal{D}_{f,h} = \underline{\mathcal{D}}_{f,h} = \sup\{w : w \in \mathcal{L}_{f,h}\}.$$

Take $w \in \mathcal{L}_{f,h}$. Let A_1 be the supremum of f. Then $w/h \leq A_1$ on Ω. Let $\varepsilon > 0$. By assumption, for μ_h-a.e. $y \in \Delta_1$ there exists a set $E_y \subset \Omega$ not minimally thin at y such that $u/h > f - \varepsilon/2$ on E_y. By the definition of $\mathcal{L}_{f,h}$ there is a neighborhood U_y of y such that $w/h < f + \varepsilon/2$ on U_y. Hence $(u - w + \varepsilon h)/h \geq 0$ on $E_y \cap U_y$, and so

$$\text{mf} \limsup_{x \to y} \frac{u - w + \varepsilon h}{h} \geq 0.$$

Observe

$$\frac{u - w + \varepsilon h}{h} \geq -A - A_1 + \varepsilon \text{ on } \Omega.$$

The extended minimum principle (Theorem 11.3.2) yields $u - w + \varepsilon h \geq 0$ on Ω. Since $\varepsilon > 0$ is arbitrary, it follows that $u \geq w$ and so $u \geq \mathcal{D}_{f,h}$ on Ω. $\square$

11.5. Minimal fine boundary limit theorem. Finally we are in a position to give the minimal fine limit theorem. The theorem will be divided to three parts, each of which deals with PWB^h solutions, Green potentials, and singular harmonic functions, respectively. Throughout this section we let h be a positive harmonic function on Ω with the representing measure μ_h.

THEOREM 11.5.1. *Let f be h-resolutive. Then*

$$\text{mf lim} \frac{1}{h} \mathcal{D}_{f,h} = f \ \mu_h\text{-a.e. on } \Delta_1.$$

PROOF. Without loss of generality we may assume that f is bounded. Let

$$f' = \text{mf limsup} \frac{\mathcal{D}_{f,h}}{h},$$
$$f^* = \max\{f, f'\}.$$

Let $u \in \mathcal{U}_{f,h}$. Observe that $\frac{u}{h} \geq -A$ on Ω and

$$\text{mf limsup} \frac{u}{h} \geq \text{mf limsup} \frac{\mathcal{D}_{f,h}}{h} = f' \text{ on } \Delta.$$

Hence

$$\text{mf limsup} \frac{u}{h} \geq f^* \text{ on } \Delta.$$

By Theorem 11.4.2 $u \geq \mathcal{D}_{f^*,h}$ on Ω. Since $u \in \mathcal{U}_{f,h}$ is arbitrary, it follows that $\mathcal{D}_{f,h}(x) \geq \mathcal{D}_{f^*,h}(x)$. By using the h-harmonic measure ν_h^x, we obtain

$$\int_\Delta f d\nu_h^x \geq \int_\Delta f^* d\nu_h^x.$$

In view of Theorem 11.4.1

$$\int_{\Delta_1} f(y) K(x,y) d\mu_h(y) \geq \int_{\Delta_1} f^*(y) K(x,y) d\mu_h(y).$$

By definition $f \leq f^*$, and hence $f(y) = f^*(y) \geq f'(y)$ μ_h-a.e. on Δ_1. Hence

$$f \geq f' = \text{mf limsup} \frac{\mathcal{D}_{f,h}}{h} \ \mu_h\text{-a.e. on } \Delta_1.$$

Similarly

$$\text{mf liminf} \frac{\mathcal{D}_{f,h}}{h} \geq f \ \mu_h\text{-a.e. on } \Delta_1.$$

Thus the theorem follows. □

THEOREM 11.5.2. *Let v be a Green potential. Then*

$$\text{mf lim} \frac{v}{h} = 0 \ \mu_h\text{-a.e. on } \Delta_1.$$

PROOF. Let $\varepsilon > 0$ and put

$$\mathcal{E} = \{y \in \Delta_1 : \text{mf limsup}_{x \to y} \frac{v(x)}{h(x)} > \varepsilon\}.$$

It is sufficient to show that $\mu_h(\mathcal{E}) = 0$. Let $E = \{x \in \Omega : v(x)/h(x) > \varepsilon\}$. Observe that mf $\text{limsup}_{x \to y} v(x)/h(x) > \varepsilon$ if and only if E is not minimally thin at y. Hence

$\mathcal{E} \subset \overline{E}^{\mathrm{mf}}$. Since $h < v/\varepsilon$ on E, it follows that $\hat{R}_h^E \leq v/\varepsilon$ on Ω. On the other hand Theorem 11.2.3 says that

$$\hat{R}_h^E(x) = \int_{\Delta_1} \hat{R}_{K_y}^E(x) d\mu_h(y) \geq \int_{\mathcal{E}} \hat{R}_{K_y}^E(x) d\mu_h(y) = \int_{\mathcal{E}} K_y(x) d\mu_h(y).$$

Thus

$$\int_{\mathcal{E}} K_y(x) d\mu_h(y) \leq \frac{v}{\varepsilon}.$$

The right hand side is a Green potential. Hence the left hand side must be equal to 0, and so $\mu_h(\mathcal{E}) = 0$. The theorem is proved. $\square$

THEOREM 11.5.3. *Let u be a singular harmonic function, i.e.,*

$$u = \int_{\Delta_1} K_y d\mu_u(y)$$

with μ_u singular with respect to μ_h. Then

$$\mathrm{mf} \lim \frac{u}{h} = 0 \ \mu_h\text{-a.e. on } \Delta_1.$$

PROOF. Let $\varepsilon > 0$ and put

$$\mathcal{E} = \{y \in \Delta_1 : \mathrm{mf} \limsup_{x \to y} \frac{u(x)}{h(x)} > \varepsilon\}.$$

It is sufficient to show that $\mu_h(\mathcal{E}) = 0$. Let $E = \{x \in \Omega : u(x)/h(x) > \varepsilon\}$. Observe that $\mathrm{mf} \limsup_{x \to y} u(x)/h(x) > \varepsilon$ if and only if E is not minimally thin at y. Hence $\mathcal{E} \subset \overline{E}^{\mathrm{mf}}$. Since $h < u/\varepsilon$ on E, it follows that $\hat{R}_h^E \leq u/\varepsilon$. On the other hand Theorem 11.2.3 says

$$\hat{R}_h^E(x) = \int_{\Delta_1} \hat{R}_{K_y}^E(x) d\mu_h(y) \geq \int_{\mathcal{E}} \hat{R}_{K_y}^E(x) d\mu_h(y) = \int_{\mathcal{E}} K_y(x) d\mu_h(y).$$

Thus

$$\int_{\mathcal{E}} K_y(x) d\mu_h(y) \leq \frac{u}{\varepsilon} = \frac{1}{\varepsilon} \int_{\mathcal{E}} K_y(x) d\mu_u(y).$$

Since μ_u and μ_h are singular with respect to each other, it follows that $\mu_h(\mathcal{E}) = 0$. The theorem is proved. $\square$

THEOREM 11.5.4. *Let u be a nonnegative superharmonic function with representing measure μ_u for the harmonic part. Then*

$$\exists \, \mathrm{mf} \lim_{x \to y} \frac{u(x)}{h(x)} = \frac{d\mu_u}{d\mu_h}(y) \quad \text{for } \mu_h\text{-a.e. } y \in \Delta_1.$$

PROOF. By the Riesz-Martin representation

$$u = v + \int_{\Delta_1} K_y d\mu_u(y)$$

$$= v + \int_{\Delta_1} K_y f d\mu_h(y) + \int_{\Delta_1} K_y d\mu_{u,s}(y),$$

where v is a Green potential, f is the Radon-Nikodym derivative $\dfrac{d\mu_u}{d\mu_h}$ and $\mu_{u,s}$ is the singular part of μ_u. In view of Theorem 11.4.1, the middle integral coincides with the PWB^h solution $\mathcal{D}_{f,h}$. Combining Theorem 11.5.1, Theorem 11.5.2 and Theorem 11.5.3, we obtain the theorem. $\square$

Bibliography

[1] D. R. Adams, *Sets and functions of finite L^p-capacity*, Indiana Univ. Math. J. **27** (1978), 611–627.

[2] _____, *Weighted nonlinear potential theory*, Trans. Amer. Math. Soc. **297** (1986), 73–94.

[3] D. R. Adams and L.-I. Hedberg, *Function Spaces and Potential Theory*, Springer, 1996.

[4] D. R. Adams and N. G. Meyers, *Bessel potentials. Inclusion relations among classes of exceptional sets*, Indiana Univ. Math. J. **22** (1973), 873–905.

[5] R. A. Adams, *Sobolev Spaces*, Academic Press, 1975.

[6] P. Ahern, *The Poisson integral of a singular measure*, Can. J. Math. **55** (1983), 735–749.

[7] H. Aikawa, *Tangential boundary behavior of Green potentials and contractive properties of L^p-capacities*, Tokyo J. Math. **9** (1986), 223–245.

[8] _____, *Comparison of L^p-capacity and Hausdorff measure*, Complex Variables **15** (1990), 223–232.

[9] _____, *Harmonic functions and Green potentials having no tangential limits*, J. London Math. Soc. **43** (1991), 125–136.

[10] _____, *Quasiadditivity of Riesz capacity*, Math. Scand. **69** (1991), 15–30.

[11] _____, *Thin sets at the boundary*, Proc. London Math. Soc. (3) **65** (1992), 357–382.

[12] _____, *Quasiadditivity of capacity and minimal thinness*, Ann. Acad. Sci. Fenn. Ser. A. I. Mathematica **18** (1993), 65–75.

[13] _____, *Integrability of superharmonic functions and subharmonic functions*, Proc. Amer. Math. Soc **120** (1994), 109–117.

[14] _____, *Densities with the mean value property for harmonic functions in a Lipschitz domain*, preprint (1995).

[15] H. Aikawa and A. A. Borichev, *Quasiadditivity and measure property of capacity and the tangential boundary behavior of harmonic functions*, Trans. Amer. Math. Soc. **348** (1996), 1013–1030.

[16] A. Ancona, *On strong barriers and an inequality of Hardy for domains in $\mathbb{R}^n$* , J. London Math. Soc. (2) **34** (1986), 274–290.

[17] D. H. Armitage, *On the global integrability of superharmonic functions in balls*, J. London Math. Soc. (2) **4** (1971), 365–373.

[18] _____, *Further result on the global integrability of superharmonic functions*, J. London Math. Soc. (2) **6** (1972), 109–121.

[19] J. P. Aubin, *Applied Functional Analysis*, Wiely, 1979.

[20] J. Bliedtner and W. Hansen, *Potential Theory, An Analytic and Probabilistic Approach to Balayage*, Springer, 1986, Universitext.

[21] M. Brelot, *On Topologies and Boundaries in Potential Theory*, Springer, 1971, Lecture Notes in Math. 175.

[22] M. Brelot and J. L. Doob, *Limites angularies et limites fines*, Ann. Inst. Fourier, Grenoble 13 (1963), 395–415.

[23] L. Carleson, *Selected Problems on Exceptional Sets*, Van Nostrand, 1967.

[24] B. E. J. Dahlberg, *A minimum principle for positive harmonic functions*, Proc. London Math. Soc. (3) 33 (1976), 238–250.

[25] J. L. Doob, *Classical Potential Theory and its Probabilistic Counterpart*, Springer, 1984.

[26] M. Essén, On Wiener conditions for minimally thin sets and rarefied sets, 41–50, Birkhäuser, 1988, pp. 41–50, Articles dedicated to A. Pfluger on the occasion of his 80th birthday.

[27] _____, *On minimal thinness, reduced functions and Green potentials*, Proc. Edinburgh Math. Soc. 36 (1992), 87–106.

[28] M. Essén and H. L. Jackson, *On the covering properties of certain exceptional sets in a half-space*, Hiroshima Math. J. 10 (1980), 233–262.

[29] C. Fefferman and E. M. Stein, H^p *spaces of several variables*, Acta Math. 129 (1972), 137–193.

[30] B. Fuglede, *Le théorème du minimax et la théorie fine du potentiel*, Ann. Inst. Fourier 15 (1965), 65–87.

[31] _____, *A simple proof that certain capacities decrease under contraction*, Hiroshima Math. J. 19 (1989), 567–573.

[32] S. J. Gardiner, *A short proof of Burdzy's theorem on the angular derivative*, Bull. London Math. Soc. 23 (1991), 575–579.

[33] K. Hansson, *Imbedding theorems of Sobolev type in potential theory*, Math. Scand. 45 (1979), 77–102.

[34] W. K. Hayman, *Subharmonic Functions, Vol. 2*, Academic Press, 1989.

[35] L. I. Hedberg and Th. H. Wolff, *Thin sets in nonlinear potential theory*, Ann. Inst. Fourier (Grenoble) 33 (1983), no. 4, 161–187.

[36] L. L. Helms, *Introduction to Potential Theory*, Wiley, 1969.

[37] R. R. Hunt and R. L. Wheeden, *Positive harmonic functions on Lipschitz domains*, Trans. Amer. Math. Soc. 147 (1970), 505–527.

[38] D. S. Jerison and C. E. Kenig, *Boundary behavior of harmonic functions in non-tangentially accessible domains*, Adv. in Math. 46 (1982), 80–147.

[39] R. Kerman and E. Sawyer, *The trace inequality and eigenvalue estimates for Schrödinger operators*, Ann. Inst. Fourier (Grenoble) 36 (1986), no. 4, 207–228.

[40] N. S. Landkof, *Foundations of Modern Potential Theory*, Springer, 1972.

[41] S. Lang, *Real Analysis, Second Edition*, Addison-Wesley, 1983.

[42] J. Lelong-Ferrand, *Étude au voisinage de la frontière des fonctions surharmoniques positive dans un demi-espace*, Ann. Sci. École Norm. Sup. 66 (1949), 125–159.

[43] F. D. Lesley, *Conformal mappings of domains satisfying a wedge condition*, Proc. Amer. Math. Soc. 93 (1985), 483–488.

[44] J. L. Lewis, *Uniformly fat sets*, Trans. Amer. Math. Soc. 308 (1988), 177–196.

[45] P. Lindqvist, *Global integrability and degenerate quasilinear elliptic equations*, J. Analyse Math. 61 (1993), 283–292.

[46] F.-Y. Maeda and N. Suzuki, *The integrability of superharmonic functions on Lipschitz domains*, Bull. London Math. Soc. **21** (1989), 270–278.

[47] M. Masumoto, *A distorsion theorem for conformal mappings with an application to subharmonic functions*, Hiroshima Math. J. **20** (1990), 341–350.

[48] _____, *Integrability of superharmonic functions on plane domains*, J. London Math. Soc. (2) **45** (1992), 62–78.

[49] _____, *Integrability of superharmonic functions on Hölder domains of the plane*, Proc. Amer. Math. Soc. **117** (1993), 1083–1088.

[50] V. G. Maz'ya, *Beurling's theorem on a minimum principle for positive harmonic functions*, Zapiski Nauchnykh Seminarov LOMI **30** (1972), 76–90, (Russian).

[51] _____, *Beurling's theorem on a minimum principle for positive harmonic functions*, J. Soviet Math. **4** (1975), 367–379, (English translation).

[52] V. G. Maz'ya, *Sobolev Spaces*, Springer, 1985.

[53] N. G. Meyers, *A theory of capacities for potentials of functions in Lebesgue classes*, Math. Scand. **26** (1970), 255–292.

[54] Y. Mizuta, *On semi-fine limits of potentials*, Analysis **2** (1982), 115–139.

[55] A. Nagel, W. Rudin, and J. H. Shapiro, *Tangential boundary behavior of functions in Dirichlet-type spaces*, Ann. of Math. **116** (1982), 331–360.

[56] A. Nagel and E. M. Stein, *On certain maximal functions and approach regions*, Adv. in Math. **54** (1984), 83–106.

[57] C. Pérez, *Two weight norm inequalities for Riesz potentials and uniform L^p-weighted Sobolev inequalities*, Indiana Univ. Math. J. **39** (1990), 31–44.

[58] P. J. Rippon, *On the boundary behaviour of Green potentials*, Proc. London Math. Soc (3) **38** (1979), 461–480.

[59] P. Sjögren, *Weak L_1 characterization of Poisson integrals, Green potentials and H^p spaces*, Trans. Amer. Math. Soc. **233** (1977), 179–196.

[60] W. Smith and D. A. Stegenga, Sobolev imbedding and integrability of harmonic functions on Hölder domains, 303–313, Walter de Gruyter, 1992, pp. 303–313, Proc. Internat. Conf. Potential Theory, Nagoya 1990 (M. Kishi, ed.).

[61] D. A. Stegenga and D. C. Ullrich, *Superharmonic functions in Hölder domains*, preprint.

[62] E. M. Stein, *Singular Integrals and Differentiability Properties of Functions*, Princeton University Press, 1970.

[63] N. Suzuki, *Nonintegrability of harmonic functions in a domain*, Japan. J. Math. **16** (1990), 269–278.

[64] _____, *Nonintegrability of superharmonic functions*, Proc. Amer. Math. Soc. **113** (1991), 113–115.

[65] _____, *Note on the integrability of superharmonic functions*, preprint (1992).

[66] A. Wannebo, *Hardy inequalities*, Proc. Amer. Math. Soc. **109** (1990), 85–95.

[67] K.-O. Widman, *Inequalities for the Green function and boundary continuity of the gradient of solutions of elliptic differential equations*, Math. Scand. **21** (1967), 17–37.

[68] P. Wojtaszczyk, *Banach Spaces for Analysts*, Cambridge University Press, 1990.

[69] K. Yosida, *Functional Analysis, Sixth Edition*, Springer, 1980.

[70] W. P. Ziemer, *Weakly Differentiable Functions*, Springer, 1989.

Index

Vol. 1540: H. Komatsu (Ed.), Functional Analysis and Related Topics, 1991. Proceedings. XXI, 413 pages. 1993.

Vol. 1541: D. A. Dawson, B. Maisonneuve, J. Spencer, Ecole d´ Eté de Probabilités de Saint-Flour XXI - 1991. Editor: P. L. Hennequin. VIII, 356 pages. 1993.

Vol. 1542: J.Fröhlich, Th.Kerler, Quantum Groups, Quantum Categories and Quantum Field Theory. VII, 431 pages. 1993.

Vol. 1543: A. L. Dontchev, T. Zolezzi, Well-Posed Optimization Problems. XII, 421 pages. 1993.

Vol. 1544: M.Schürmann, White Noise on Bialgebras. VII, 146 pages. 1993.

Vol. 1545: J. Morgan, K. O'Grady, Differential Topology of Complex Surfaces. VIII, 224 pages. 1993.

Vol. 1546: V. V. Kalashnikov, V. M. Zolotarev (Eds.), Stability Problems for Stochastic Models. Proceedings, 1991. VIII, 229 pages. 1993.

Vol. 1547: P. Harmand, D. Werner, W. Werner, M-ideals in Banach Spaces and Banach Algebras. VIII, 387 pages. 1993.

Vol. 1548: T. Urabe, Dynkin Graphs and Quadrilateral Singularities. VI, 233 pages. 1993.

Vol. 1549: G. Vainikko, Multidimensional Weakly Singular Integral Equations. XI, 159 pages. 1993.

Vol. 1550: A. A. Gonchar, E. B. Saff (Eds.), Methods of Approximation Theory in Complex Analysis and Mathematical Physics IV, 222 pages. 1993.

Vol. 1551: L. Arkeryd, P. L. Lions, P.A. Markowich, S R S. Varadhan. Nonequilibrium Problems in Many-Particle Systems. Montecatini, 1992. Editors: C. Cercignani, M. Pulvirenti. VII, 158 pages 1993.

Vol. 1552: J. Hilgert, K.-H. Neeb, Lie Semigroups and their Applications. XII, 315 pages. 1993.

Vol. 1553: J.-L- Colliot-Thélène, J. Kato, P. Vojta. Arithmetic Algebraic Geometry. Trento, 1991. Editor: E. Ballico. VII, 223 pages. 1993.

Vol. 1554: A. K. Lenstra, H. W. Lenstra, Jr. (Eds.), The Development of the Number Field Sieve. VIII, 131 pages. 1993.

Vol. 1555: O. Liess, Conical Refraction and Higher Microlocalization. X, 389 pages. 1993.

Vol. 1556: S. B. Kuksin, Nearly Integrable Infinite-Dimensional Hamiltonian Systems. XXVII, 101 pages. 1993.

Vol. 1557: J. Azéma, P. A. Meyer, M. Yor (Eds.), Séminaire de Probabilités XXVII. VI, 327 pages. 1993.

Vol. 1558: T. J. Bridges, J. E. Furter, Singularity Theory and Equivariant Symplectic Maps. VI, 226 pages. 1993.

Vol. 1559: V. G. Sprindžuk, Classical Diophantine Equations. XII, 228 pages. 1993.

Vol. 1560: T. Bartsch, Topological Methods for Variational Problems with Symmetries. X, 152 pages. 1993.

Vol. 1561: I. S. Molchanov, Limit Theorems for Unions of Random Closed Sets. X, 157 pages. 1993.

Vol. 1562: G. Harder, Eisensteinkohomologie und die Konstruktion gemischter Motive. XX, 184 pages. 1993.

Vol. 1563: E. Fabes, M. Fukushima, L. Gross, C. Kenig, M. Röckner, D. W. Stroock, Dirichlet Forms. Varenna, 1992. Editors: G. Dell'Antonio, U. Mosco. VII, 245 pages. 1993.

Vol. 1564: J. Jorgenson, S. Lang, Basic Analysis of Regularized Series and Products. IX, 122 pages. 1993.

Vol. 1565: L. Boutet de Monvel, C. De Concini, C. Procesi, P. Schapira, M. Vergne. D-modules, Representation Theory, and Quantum Groups. Venezia, 1992. Editors: G. Zampieri, A. D'Agnolo. VII, 217 pages. 1993.

Vol. 1566: B. Edixhoven, J.-H. Evertse (Eds.), Diophantine Approximation and Abelian Varieties. XIII, 127 pages. 1993.

Vol. 1567: R. L. Dobrushin, S. Kusuoka. Statistical Mechanics and Fractals. VII, 98 pages. 1993.

Vol. 1568: F. Weisz, Martingale Hardy Spaces and their Application in Fourier Analysis. VIII, 217 pages. 1994.

Vol. 1569: V. Totik, Weighted Approximation with Varying Weight. VI, 117 pages. 1994.

Vol. 1570: R. deLaubenfels, Existence Families, Functional Calculi and Evolution Equations. XV, 234 pages. 1994.

Vol. 1571: S. Yu. Pilyugin, The Space of Dynamical Systems with the C⁰-Topology. X, 188 pages. 1994.

Vol. 1572: L. Göttsche, Hilbert Schemes of Zero-Dimensional Subschemes of Smooth Varieties. IX, 196 pages. 1994.

Vol. 1573: V. P. Havin, N. K. Nikolski (Eds.), Linear and Complex Analysis – Problem Book 3 – Part I. XXII, 489 pages. 1994.

Vol. 1574: V. P. Havin, N. K. Nikolski (Eds.), Linear and Complex Analysis – Problem Book 3 – Part II. XXII, 507 pages. 1994.

Vol. 1575: M. Mitrea, Clifford Wavelets, Singular Integrals, and Hardy Spaces. XI, 116 pages. 1994.

Vol. 1576: K. Kitahara, Spaces of Approximating Functions with Haar-Like Conditions. X, 110 pages. 1994.

Vol. 1577: N. Obata, White Noise Calculus and Fock Space. X, 183 pages. 1994.

Vol. 1578: J. Bernstein, V. Lunts, Equivariant Sheaves and Functors. V, 139 pages. 1994.

Vol. 1579: N. Kazamaki, Continuous Exponential Martingales and BMO. VII, 91 pages. 1994.

Vol. 1580: M. Milman, Extrapolation and Optimal Decompositions with Applications to Analysis. XI, 161 pages. 1994.

Vol. 1581: D. Bakry, R. D. Gill, S. A. Molchanov, Lectures on Probability Theory. Editor: P. Bernard. VIII, 420 pages. 1994.

Vol. 1582: W. Balser, From Divergent Power Series to Analytic Functions. X, 108 pages. 1994.

Vol. 1583: J. Azéma, P. A. Meyer, M. Yor (Eds.), Séminaire de Probabilités XXVIII. VI, 334 pages. 1994.

Vol. 1584: M. Brokate. N. Kenmochi, I. Müller, J. F. Rodriguez, C. Verdi, Phase Transitions and Hysteresis. Montecatini Terme, 1993. Editor: A. Visintin. VII. 291 pages. 1994.

Vol. 1585: G. Frey (Ed.), On Artin's Conjecture for Odd 2-dimensional Representations. VIII, 148 pages. 1994.

Vol. 1586: R. Nillsen, Difference Spaces and Invariant Linear Forms. XII, 186 pages. 1994.

Vol. 1587: N. Xi, Representations of Affine Hecke Algebras. VIII, 137 pages. 1994.